魅惑夜生活 The Charming Night 目录

CONTENTS

魅惑·酒吧 Fascinating Bar

Charming Nightclub

The dazzling night life could make people think about nightclub, bar, Disco...
yes, that's them, which came from western country in the 1980s, they have been popular since the speedy economic development of Hongkong, and these entertainment spaces extended their influence to Guangdong after Hong Kong economic crisis. Within Guangdong province, especially Shenzhen and Dongguan, these kinds of night life entertainments are prevailing. Until today, the investors across the country also like to choose Shenzhen, where designers could make an investigation about the design from the business pattern to the furnishing style, and the advantageous geographic condition of Shenzhen benefits such entertainment space design.

As time changes, the business mode for nightclub is one thing, and the furnishing is and the other thing. The updating of the design style is also fast, and the design style almost takes a change within three years or one year. Compared with wineshop, it has lots of profits. The high-tech materials benefits the updating of nightclub design and the designer has more selections and requirements for lighting and materials.

In the past, the nightclub design style is mainly based on extravagant style, but as the changing ages, some nightclub players are fed up with such design style. They ask for higher design levels, and the design itself should also have its own culture, with less lighting installation yet have the nightclub flavor, the top level nightclub turns to be the business nightclub. However, it has something to do with the culture of each city. The night life is all about entertainment, so the design of nightclub could be flexible. Dear designers, we hope you could have more imaginative design for nightclub design…

魅惑·夜店

提到魅惑夜生活，人们应该很快会联想到夜总会、酒吧、DISCO……没错，就是它们，自20世纪80年代由西方国家传入中国，在香港经济的起飞、发展时期甚为流行，90年代受香港经济影响后渐往澳门及广东等地区发展，其中在深圳、东莞等周边地区极为盛行，直到经济高速发展的今日，中国不同地方的投资商、设计师还是喜欢前往深圳考察，从经营的模式到装修的风格，或许是因为它理想的地理位置给了它优越的条件。

夜店随着时间在不停的变化，经营的模式是其一，而装修的感觉是其二，风格的更新速度极快，折旧也很快，基本上每年一小装三年一大装，当然这也因为它有着暴利的经济基础条件，这跟酒店是不一样的。高科技材料的出现使得夜店设计日新月异，在对灯光及材料的使用上有了更多的选择与要求。

过去的夜店多以灯红酒绿、纸醉金迷为主打，随着时代的变迁，一些常在夜店的玩家们有些开始对这些感觉腻了，认为是过时的、没有文化内涵的装修，开始提出更高的要求，要够有档次，但还得装修得有文化气质，灯光不想太花太暗，却又要有夜店的味道，于是现在的高端夜店都开始向商务型转变。当然这也跟不同城市的文化差异有关，一个刚进入夜店文化不久的群体马上进入一个新的夜店文化是难以接受的，夜店也是如此，因人而异。但有一点是共同的，夜生活就是娱乐，既然是娱乐，那夜店设计就可以没有章法。来吧，各位设计师朋友，期待大家更多的天马行空……

辛军　文

普利维行星好莱坞酒店&拉斯维加斯赌场

'PRIVE' Planet Hollywood Hotel and Casino Las Vegas

设计师：Francois Frossard
设计公司：FFD Inc
项目地点：美国拉斯维加斯
项目面积：1116平方米

Upon entering the venue, guests will be met by Prive's towering main bar. With a concave-shape granite top, backlit tortoise shell façade featuring high gloss millwork, and a whimsical inverse pyramid shaped crystal chandelier, the bar will make its own stellar design statement. To the right, demure VIP's will enjoy a cozy, private rotund seating area for intimate conversations and low-key mingling, featuring its own private bathrooms. Featuring 12,000 square feet of luxuriously appointed design elements created by renown designer Francois Frossard, Prive mixes the ethereal whimsy of South Beach with dramatic go-go elements that are quintessential. In addition to a distinct atmosphere, guests can expect surprise performances, celebrity guests and world-class service.

魅惑

The bizarre design style doesn't make people feel uncomfortable, instead for visual sense, it is more powerful.

怪诞的风格丝毫不会让人们感到不舒适，在视觉表现上，显得更具力量感。

privé

进入大厅，迎面而来的是普利维的主酒吧，以凹形花岗岩造型为顶部，背光的龟壳面来突出高光的木制装饰，还有一个倒金字塔似的水晶吊灯，整个酒吧有其各自的设计特色。在酒吧的右面，为了方便顾客之间进行亲密的交流，贵宾区设有舒适的、隐秘的专座，并带有独立的卫生间。

在这个约1100平方米的奢华空间里，所有的翻新工作都由Francois Frossard亲自设计。普利维融合了South Beach的怪诞风格，同时也融入了具有戏剧化效果的典型拉斯维加斯go-go元素。在这里除了享受酒吧独特的氛围之外，顾客还可以观看到令人称奇的表演，偶遇世界名流和享受一流的服务。

privé
las vegas

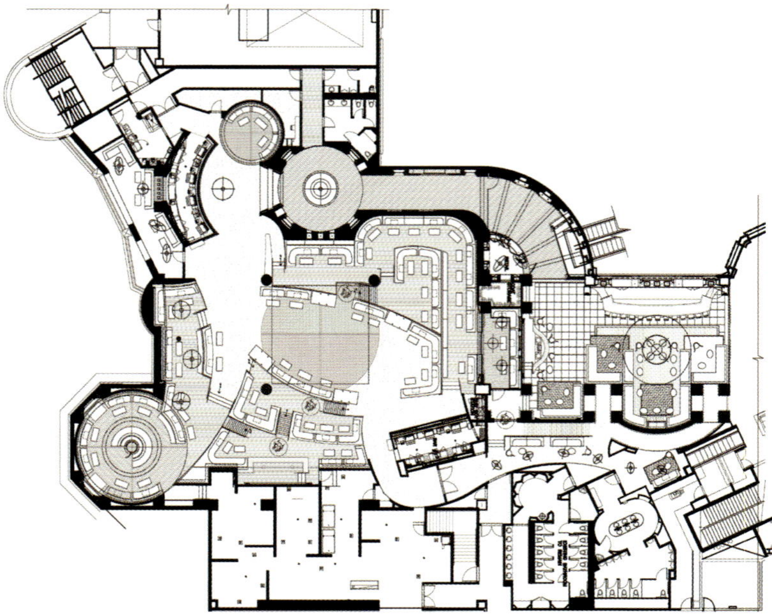

钱隆商务量贩式KTV

Qianlong Business KTV

设计师：霍庆涛
设计公司：大石代设计咨询有限公司
项目地点：中国石家庄
项目面积：3000平方米
主要材料：地砖、壁纸、木饰面、不锈钢等
项目摄影：邢振涛

The KTV engaged mainly in business consumption provides customers with private rooms for entertainment as well as top quality buffet and notch recording studio. As early as the conceiving phase, the designer has explored the means and space of entertainment really belonging to our Chinese, who thought of traditional opera, tune cards and landscape paintings made by scholars to express their feelings. After repeated deliberation, the designer finally adopted the ideas of "scene setting". At least 30 large or small scenic spots are arranged at public spaces and private rooms, among which there're 24 main views called "Qianlong 24 scenic spots". Each scenic spot is given a meaningful lucky name, such as "the lotus pool by moonlight" at the entrance hall, "bearing both flowers and fruits" at the cashier counter, "round as pearls and smooth as jade", "shining splendor in succession" , "quiet dream behind a curtain" along the hallways and "flower like youth" at the cafeteria. Using these scenes, the designer infuses numerous elements of China's humanity complex, history and culture into the entertainment space, giving it the soul and energy and making people learn about China and culture while entertaining themselves.

魅惑

Combining the Chinese culture and history, the design affords the soul and life into the space.

将中国的人文情结、历史、文化融入娱乐空间中，赋予空间灵魂和生命。

化妆间

K50

K35

风味串烧

K26
K25
K23

K17
K19
K20

本项目是以商务消费为主的量贩式KTV，在给顾客提供娱乐包间的同时，还为顾客提供高品质的自助餐和一流的录音棚。在本案的创意阶段，设计师就在探索设计真正属于中国人自己的一种娱乐方式和娱乐空间。设计师联想到了戏曲，想到了调牌，想到了文人笔下用来抒发感情的山水画，经过反复推敲，最终采用了“置景式”这一设计思路。在公共空间和包间，设计师安排了大大小小不下30个景观点，主景观有24个，称之为“乾隆二十四景”，每一个景观都有吉祥寓意的名称：大厅的“荷塘月色”，收款台的“花实同生”，走廊的“珠圆玉润”、“节节生辉”、“一帘幽梦”，自助餐的“花样年华”等。设计师利用这些景观，将中国的人文情结、历史、文化等诸多元素注入这一娱乐空间当中，为空间注入灵魂，注入活力，让人们在娱乐身心的同时，感知中国，感知文化。

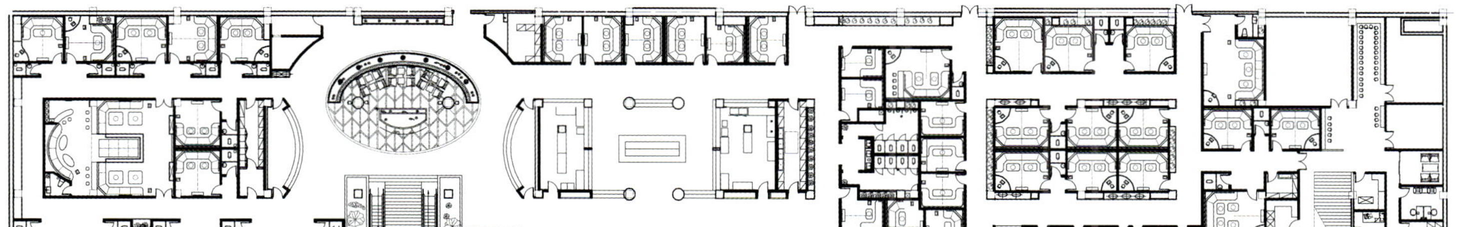

音悦汇量贩式KTV

Yinyuehui KTV

设计师：霍庆涛、丁洁
设计公司：大石代设计咨询有限公司
项目地点：中国沙河
项目面积：2100平方米
主要材料：白色瓷砖、彩色镜面、壁纸等
项目摄影：邢振涛

For this entertainment space designed with the theme of colors, the designer seeks to use the most concise lines and words to achieve the most varied spatial effect. The first and second floors are defined as red and white tone to make the space full of warmth and energy, where the young people would release their passions. The third floor is set as white and green tone, which is natural, fresh and suitable for family reunion and birthday party. The fourth floor based on black and white tone appears sophisticated, elegant, which is only available to business people. The same styling but different colors bring people different feelings of space and multiple choices.
"Bionics" is utilized in the design of space styling, which is inspired by "butterfly wings", "gene chain", "honeycomb" , "cell" and mineral crystals, not only enriching the spatial looks, layers but also increasing the content and meaning of the space. "A line sky" designed at the roof of corridor well settles the problem of corridor's lower height. Responding to each other, roof mirrors and colorful mirrors on the wall interact with the people walking through the corridor to the effect as if people reach an illusory realm and everything seems drifting from place to place while existing in the reality.

魅惑

The abundant colors show us the theme of the space while bring the visual enjoyment to the customers.

丰富的色彩展示出了空间的主题，从而也给顾客带来丰富的视觉享受。

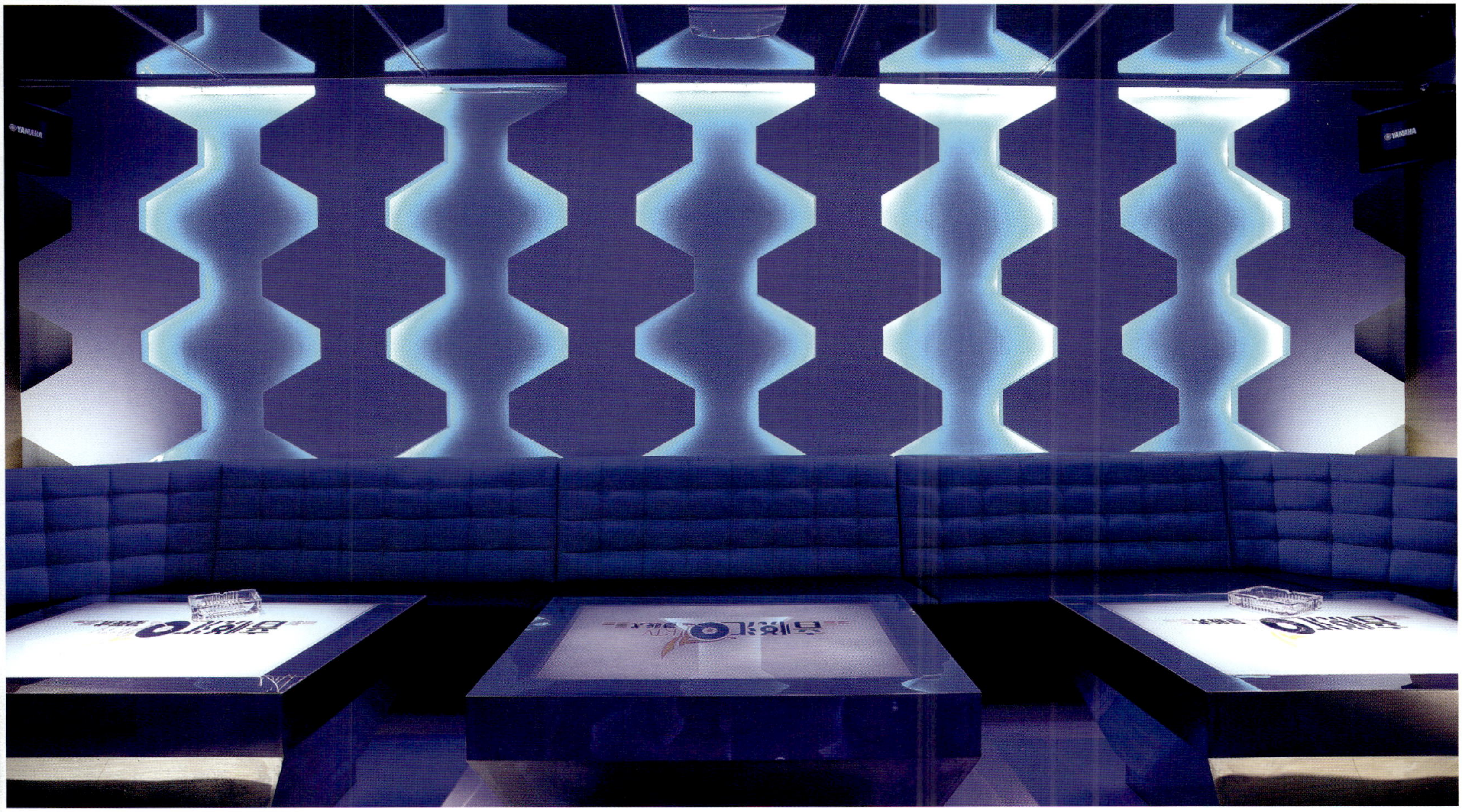

203

洗手间
TOILET
洗手间

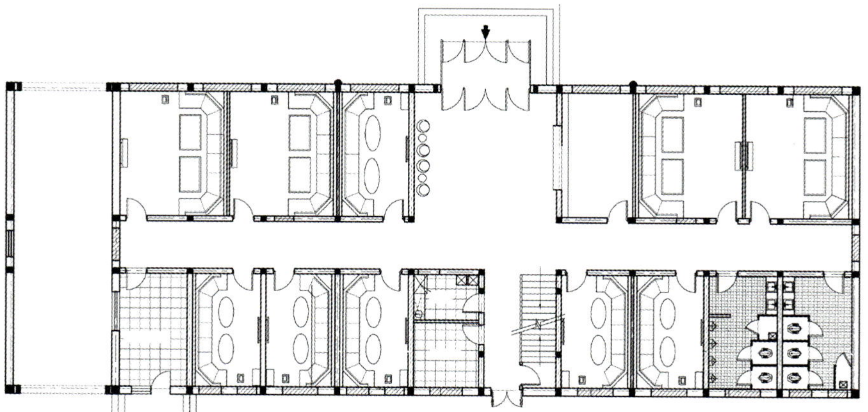

一层

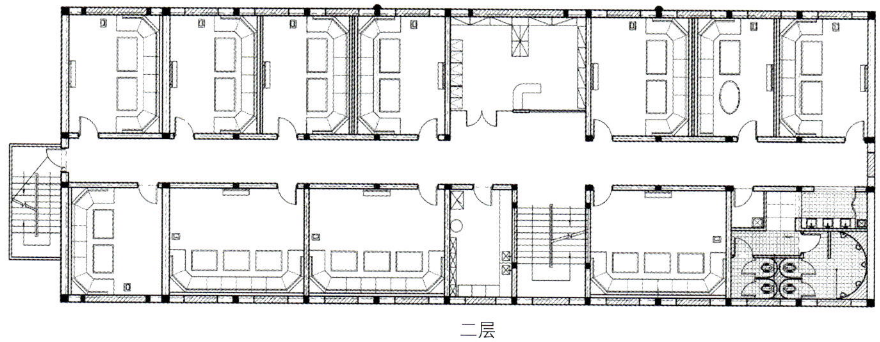

二层

这是一个以色彩为主题设计的娱乐空间，设计师力求在设计中运用最简洁的线条和笔墨达到最丰富的空间效果。设计师将首层和二层空间定义为红白色调，使它富有热情和活力，成为年轻一族释放激情的地方。三层为白绿调，自然、清新，适合家庭聚会和生日聚会。四层为黑白色调，高贵、典雅，专门为商务人群设置。相同的造型，不一样的色彩，带给人们不同的空间感受和多重的选择。

在空间的造型设计上运用了“仿生学”的设计，从“蝴蝶翅膀”、“基因链”、“蜂巢”、“细胞”以及矿物质的结晶中汲取灵感，既丰富了空间造型和层次，又增加了空间的内涵和意蕴。走廊顶部“一线天”的设计，很好地解决了走廊层高过低的问题，顶部镜面和墙面彩镜交相呼应，与穿行于走廊之中的人群形成互动，使人的身心达到一种“虚无”的境界，使一切变得飘忽不定却又真实存在。

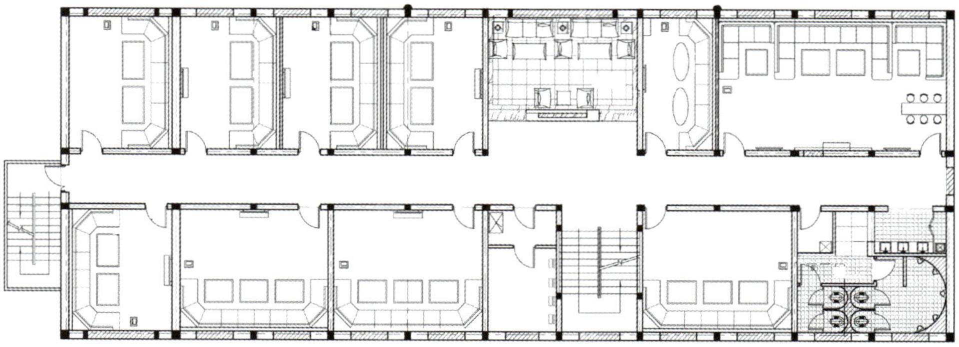

三层

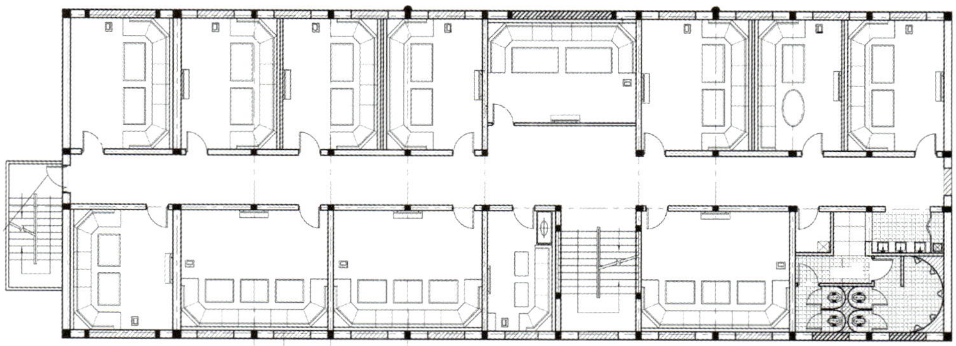

四层

爱伊美娱乐商务会所

Aiyimei Entertainment Business Club

设 计 师：程礼波
设计公司：宁波九间堂装饰工程设计有限公司
项目地点：中国奉化
项目面积：2000平方米
主要材料：进口玉木纹、进口木纹黄大理石、造型钛金板、成品装饰面板、白色艺术条砖等

Located on the third floor of four-star Aiyimei Hotel in Fenghua, this project is conceived to create a unique, sophisticated, modern business club leading the fashionable trend.Breaking through the usual design techniques, the hall is deigned with an all clear, open layout, enhancing the interactive happiness to the highest. The primary and secondary corridors, made of different materials and designed with complex, overlapped geometric patterns, bring an absolutely outstanding shock visually and sensationally to the guests with its sense of multi-layers, spatial extension and lighting variance in a more rhythmical way. Furthermore, the private rooms with an even out-of -box design, are used with diversified materials to promise every consumer safety, health, delight and fashion, which is considered as the starting point of the design. The rooms are designed to express other sensual pleasure in addition to the sense of hearing through an absolutely overturning method. Joyful, passionate and visual impact stimulating sense organs is the main inspiration or soul running through the design. The glamour of the project lies in its bold, innovative concept, ingenious ideas not sticking to one pattern and this is the biggest "playful design" given by the owner to the design team.

魅 Not lacking of elegance and fashion, this is a business club.

惑 不乏精致，不乏时尚，也不乏潮流的商务休闲会所。

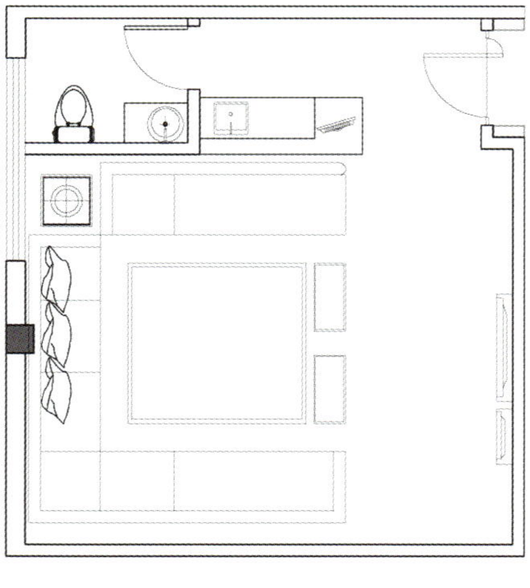

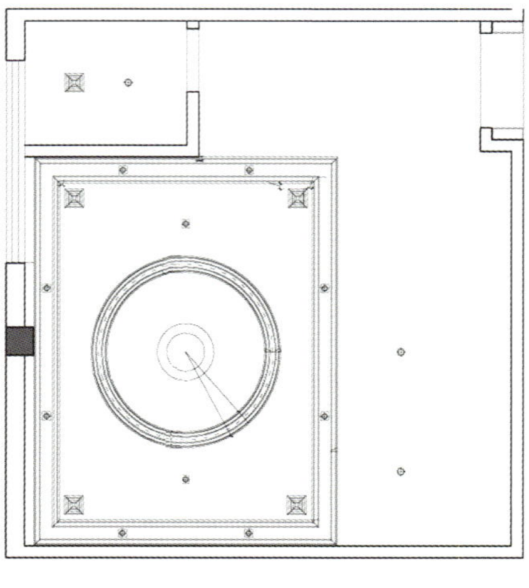

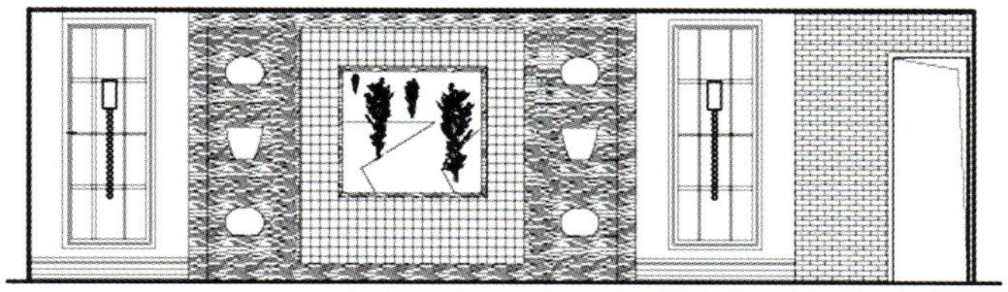
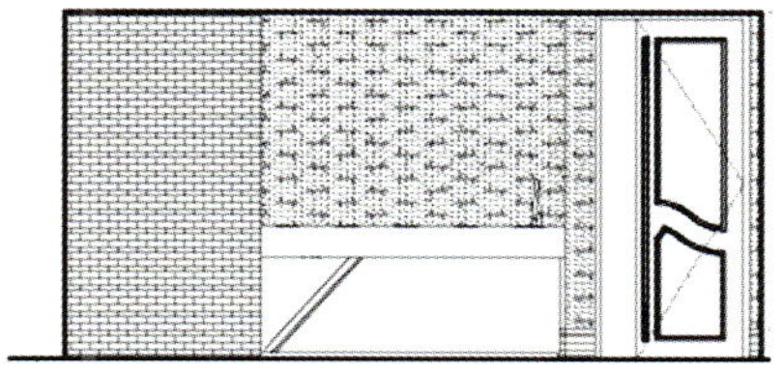
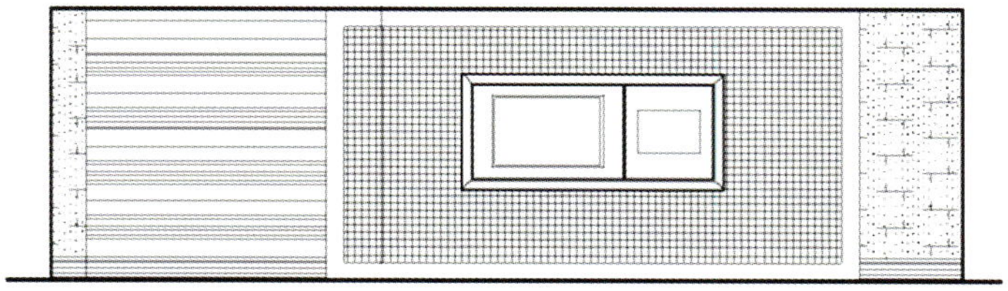
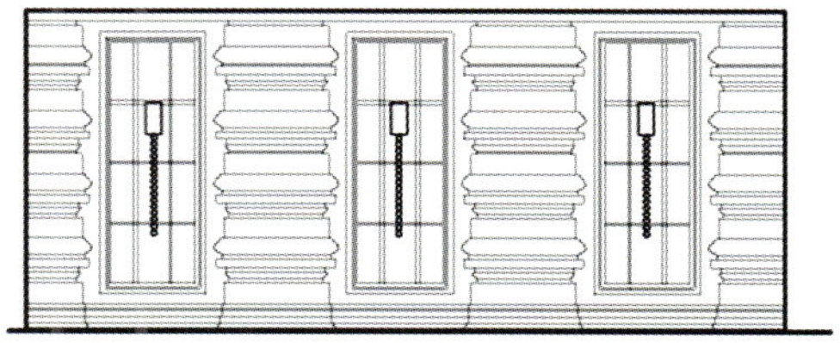

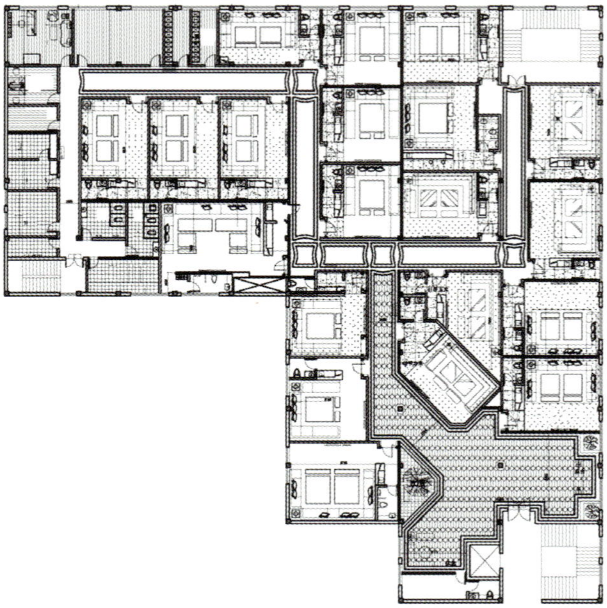

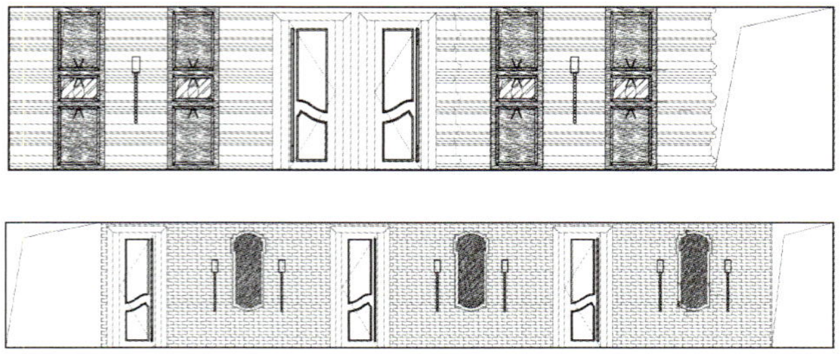

爱伊美娱乐商务会所位于中国奉化市四星级爱伊美大酒店三层，本案的设计理念是营造一个独特、精致、充满现代气息，并能引领时尚潮流的商务会所。大厅的格局，打破以往的设计手法，采用全透明开放形式，将互动的欢乐提升至最高点。主次走道采用多种材料元素和复合叠加的几何设计理念，以其多重层次感、空间延伸感及灯效变化感，为客人呈现别样的视觉及感官冲击。包厢内的设计更是突破传统形式，以多样的材质变化为每一位消费者打造、安全、健康、欢乐和时尚的娱乐空间，这也是设计的起点，包厢设计以绝对的颠覆手法表现了除听觉以外的感官享受。欢愉、热情、视觉的趣味、感官冲击是贯穿本案设计的主要灵感，更是灵魂。本案的魅力在于它大胆创新的设计理念，不拘一格的奇思妙想，这也是本案业主赋予设计团队们最大的“玩乐设计”。

新奥尔良赌场娱乐城

Masquerade at Harrah's New Orleans

设计公司：Watg 设计公司
合作设计师：Montgomery Roth
项目地点：美国新奥尔良
项目面积：930平方米

This is a Baroque style design. In New Orleans, people pursues a kind of fashionable and retro combined art. The classicist thinks that Baroque is a kind of degraded art, and until now the public has a righteous assessment. This project pursues luxurious design style which keeps the hedonism design style. The important feeling which masquerade gives to people is enthusiasm, and you could find it here. The enthusiasm breaks the quiet and harmonious atmosphere, and it is with strong romantic flavor. The stage affords a relaxing place for people, and they could dance to the music.

魅惑

The artistic feeling is throughout the space while the design combines fashionable and antique flavor.

用艺术贯穿空间，将时尚与复古结合。

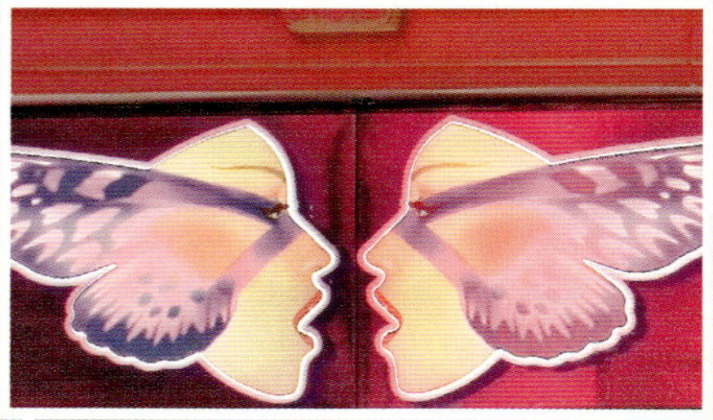

在空间的设计上，设计师带来了巴洛克的风格。在新奥尔良，人们追求的是一种时尚和复古相互结合的艺术。古典主义者认为巴洛克是一种堕落瓦解的艺术，只是到了后来，才对巴洛克艺术有了一个较为公正的评价。本案坚持豪华的特色，既有宗教的特色又有享乐主义的色彩。娱乐城最重要的是要带给人激情，这点在这里得到了实施，它打破理性的宁静和谐，具有浓郁的浪漫主义色彩。舞台成为了一个可以宣泄情绪的地方，人们可以伴随着音乐的节奏在这里跳舞。

美高美酒店休闲中心

Meigaomei Recreational Center

设计师：林永锋、姜丽玲、王密、姜婷婷、黄新媛等
设计公司：极度空间酒店设计机构
项目地点：中国东莞
项目面积：8000平方米
主要材料：花瓣、编织物、透光大理石、泰纹地砖、玫瑰韩国壁纸、艺术玻璃、透光大理石等

The recreational center of the hotel reflects the taste of the fashion and leisure, creating a lot of private space, which provide the KTV room and billiards room for the guests to relax and enjoy in.

The broad leisure center lobby is presented to the customers in a simple and elegant way without any uncomfortable feelings. The warm colors are still the first choice of the color design, which is dotted with blue, gray and other colors at the same time, addressing the layer of the colors. The perfect space shows its changing characteristics. The staggered spacial lines achieved a good fashion result.

魅惑

The spacious space creates the relaxing entertainment center with best quality, and the replete color makes people love it at first sight.

开阔的视野呈现出了休闲娱乐中心的最佳品质。饱满的色彩让人一见倾心。

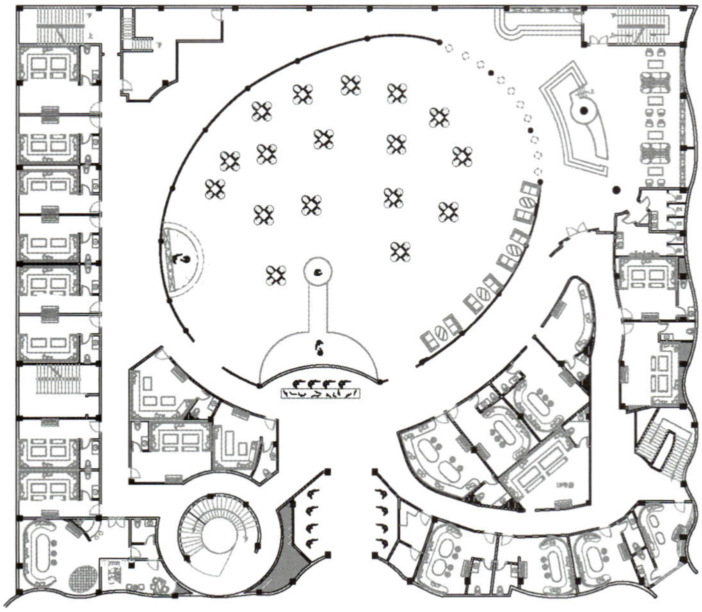

一层

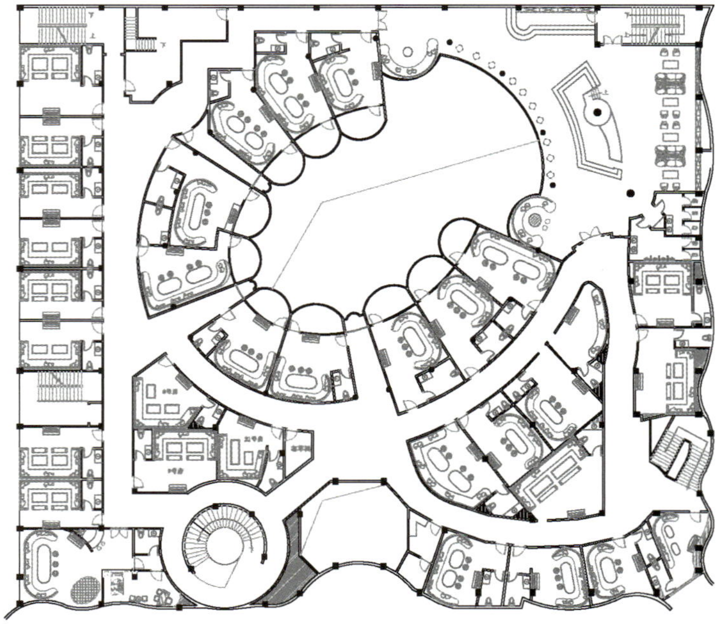

二层

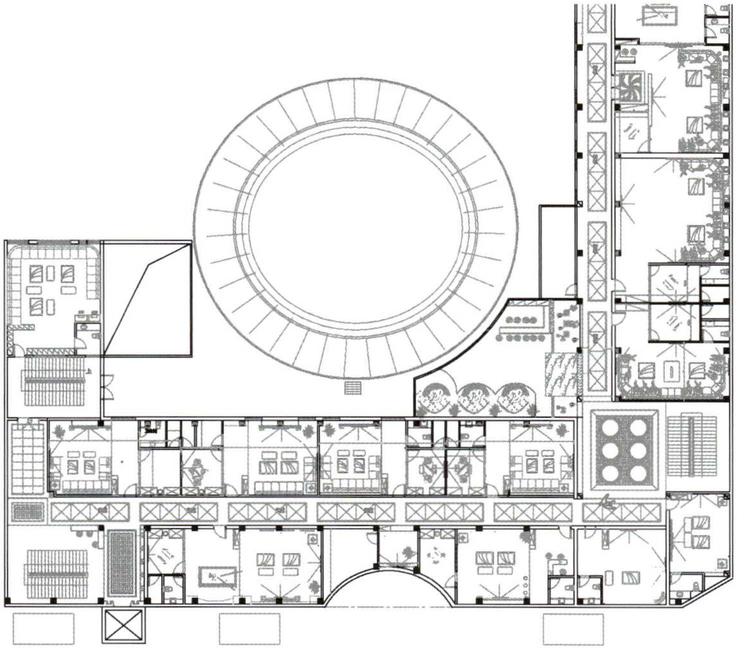

三层

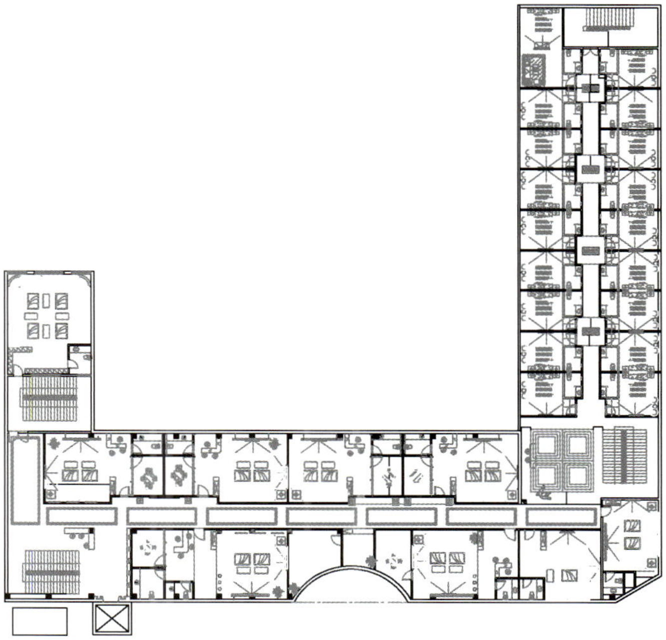

四层

该酒店休闲中心的设计体现了时尚与休闲的味道。休闲中心打造了很多包间， 在此朋友之间可以三五成群地唱歌或者享受其他放松休闲活动，如台球室为喜爱台球的客人创造了良好的环境。

休闲中心的大堂以开阔的姿态呈现在顾客面前，大堂的设计简洁大方，不受任何约束，这让顾客倍感舒适。色彩上，使用了暖色调，同时搭配蓝色、灰色等，突出空间的层次感。完美的空间造型呈现出多变的特点，空间线条之间的交错达到了很好的时尚效果。

乌鲁木齐IN BAR俱乐部

Urumchi IN BAR Club

设计师：孙亮
设计公司：成都优佰文化传播有限公司
项目地点：中国乌鲁木齐
项目面积：800平方米
主要材料：实木、灰镜、墙纸、石材、布艺等

In this project, the designer abandons normal arrangement, using different features of space arrangements and making the layout full of interests and unique characters while promising the space practicability. The designer made the clear division between people and guests, which could make the guests indulge themselves into the entertainment atmosphere.

As for the decorative materials, the designer doesn't choose the expensive materials. He uses normal materials to decorate the formations and colors. The designer has the special requirement for lighting, fully using cold and warm lighting to satisfy the different demands for guests.

魅惑

Abandoning the normal principles while promising the practicability of the space, the designer wants to express more unique characteristics and interests of the space.

摈弃常规，在保证空间实用性的基础上，表达更多的个性和空间趣味。

InBar

FASHION CLUB

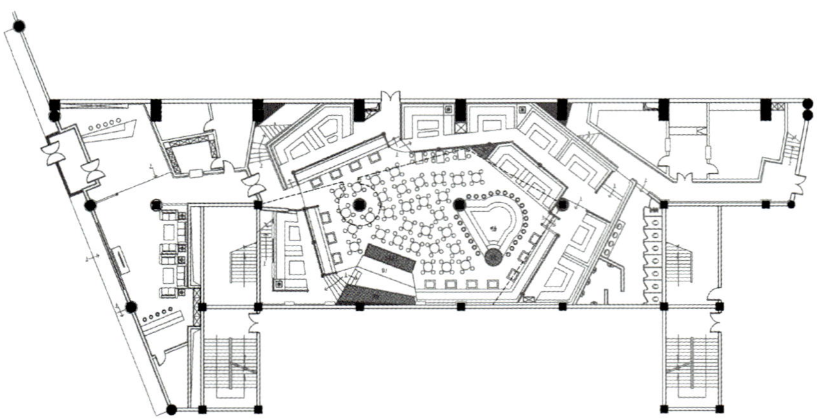

一层

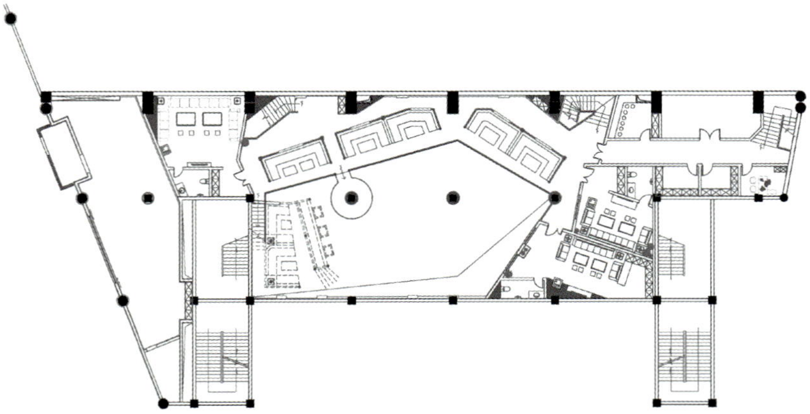

二层

InBar

JAGUAR
InBar
FASHIONCLUB

在本案中，设计师摒弃了常规的布局方式，充分利用空间布局的特点，把平面布局做得充满趣味、个性鲜明，同时也尽可能地保证了空间的实用性。特别是对顾客和服务人员的动线走向做出了区分，让顾客能够充分地融入到俱乐部的娱乐氛围当中。

在装饰材料上，设计师并没有选用特别昂贵的材料，而是用常规的材料，在造型和色彩上做出调整。另外，设计师特意对灯光提出了特别的要求，充分利用冷暖色灯光的配合，满足不同区域消费群体的需求。

布谷鸟俱乐部
The Cuckoo Club

设计师：Tim Mutton
设计公司：Blacksheep 设计公司
项目地点：英国伦敦

The Cuckoo Club, a nightclub and destination restaurant, owned by Unicorn City Ltd and designed by Blacksheep, has opened on the former site of the Stork Rooms in Swallow St, London W1. Once a revered staple of London's nightlife, favoured by bohemians and aristocracy alike.

The Cuckoo Club offers high-octane rock'n'roll glamour, with an overriding accent on comfort, fun and service. Although entry will not be constricted to members only, the launch committee's honorary members - invited figures from the worlds of music, fashion, film, art, photography, high society, food and journalism - will subtly set the tone for the club.

The interior configuration has an emphasis on flexibility of use, with easily changeable furniture and feature areas that can also be used for staging private parties or promotional events. "We envisaged the Cuckoo Club as a kind of filmset", explained Tim Mutton, "creating a grand, dramatic, fantasy house with huge doors, sweeping stairs, silk and voile drapes, endlessly reconfigurable LED lighting and lots of fun, statement design features".

魅惑

This project has flexible and dramatic space feeling which benefits from the world distinguished and superior service, and the space mainstream quality is highlighted by this way.

布置灵活且具有戏剧感的大空间，得益于世界名流的基调以及优质的服务，空间的主流品质得以提升。

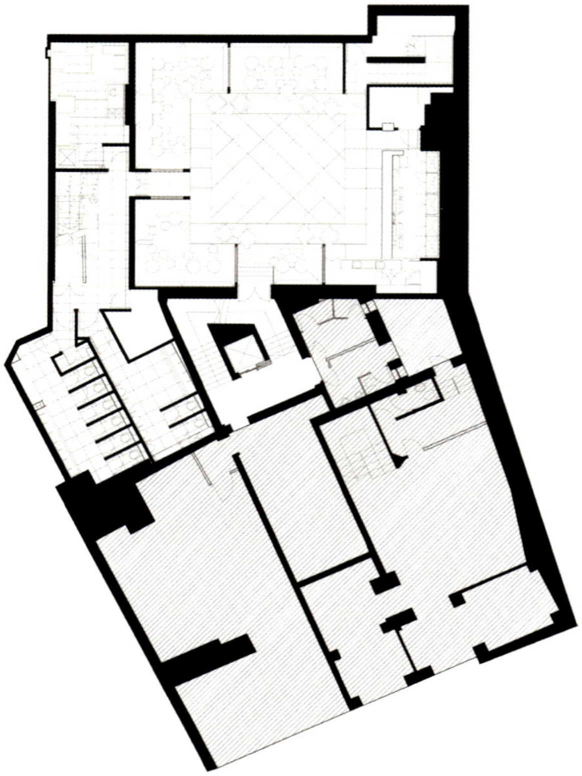

负一层

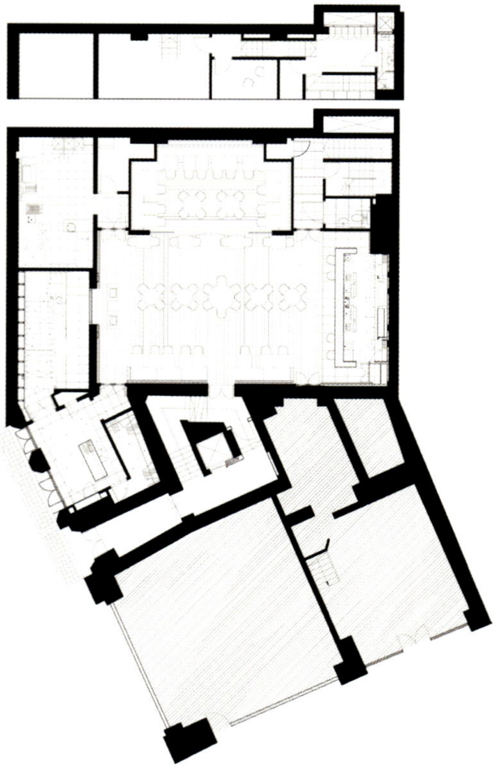

一层

布谷鸟俱乐部是一个集夜店与餐厅于一体的俱乐部，业主为Unicorn City公司，由Blacksheep设计。伦敦的夜生活曾是非常受中产阶级和王公贵族推崇的。

布谷鸟俱乐部为人们提供了一种极炫的摇滚风，主要强调舒适、有趣以及优质的服务。俱乐部也并非专为会员所开，还邀请世界著名的音乐、电影、艺术、摄影、媒体等时尚界人士，这些主流人群为这个俱乐部定下了主要的基调。

室内的布置注重其灵活性，用可移动的家具以及充满特色的区域围合成私人俱乐部和集会区。“我们把布谷鸟俱乐部当成电影布景来设计，” Tim Mutton说道，“营造了一种豪华的、充满戏剧性与幻想的房子，并且有着巨大的门框、楼梯、薄纱窗帘、无数的LED聚光灯，同时也有许多乐趣、特色在其中。”

蓝黛CLUB

Blue and Black CLUB

设 计 师：郎俊峰
设计公司：河南柏言装饰工程有限公司
项目地点：中国漯河
项目面积：1360平方米
主要材料：金属板、水晶板、亚克力板、皮包块、艺术玻璃、蓝钻石材、LED灯、雕花板等
项目摄影：历柏言

As for entertainment place, style is a layout, and it is also the expression of a taste. As the saying goes, it's the design that determines success or failure and the style that shapes future. The designer acutely grasped the market's demands to carry out a bold interior design, in which a large number of decoration materials with reflective surface are used, for example, dark vitrified tiles, black glass, grey glass are matched with hanging curtain, lED lamps, patterned wallpaper, fabric sofa and light-emitting glass coffee table. Through the designer's ingenious ideas, materials combined by the means of colors, patterns and lights are unified and enhanced in the terms of the style, which, under the reflection between LED and glass mirrors, highlight the magnificent, chic and romantic Oriental Entertainment City and showcase the beauty of a leisurely space integrating cultural information and entertainment function.

魅惑

Based on design market, the designer has exquisite design methods and bold selection which create popular space to us.

以市场为出发点，巧妙构思，大胆选材，营造出了一处供大众休闲的空间。

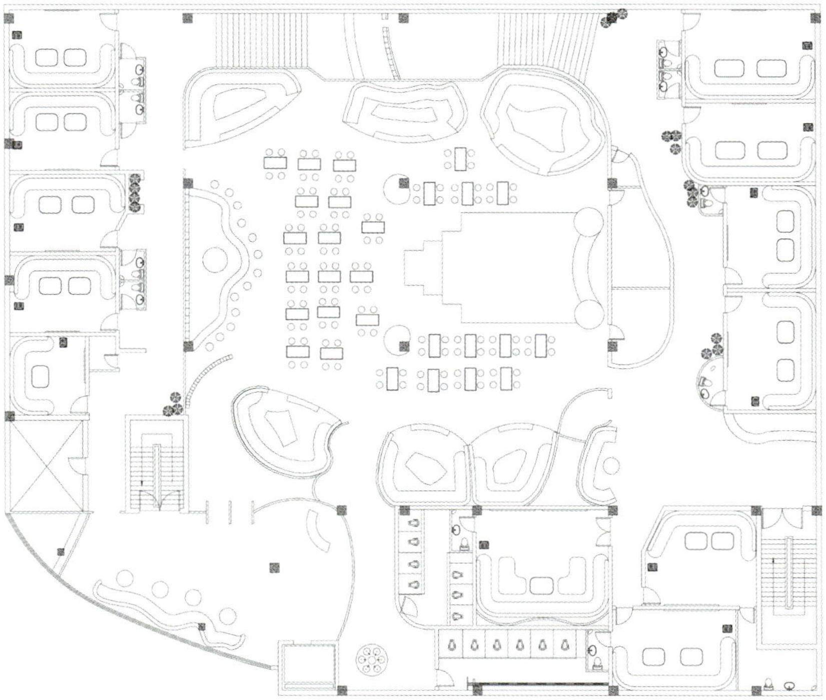

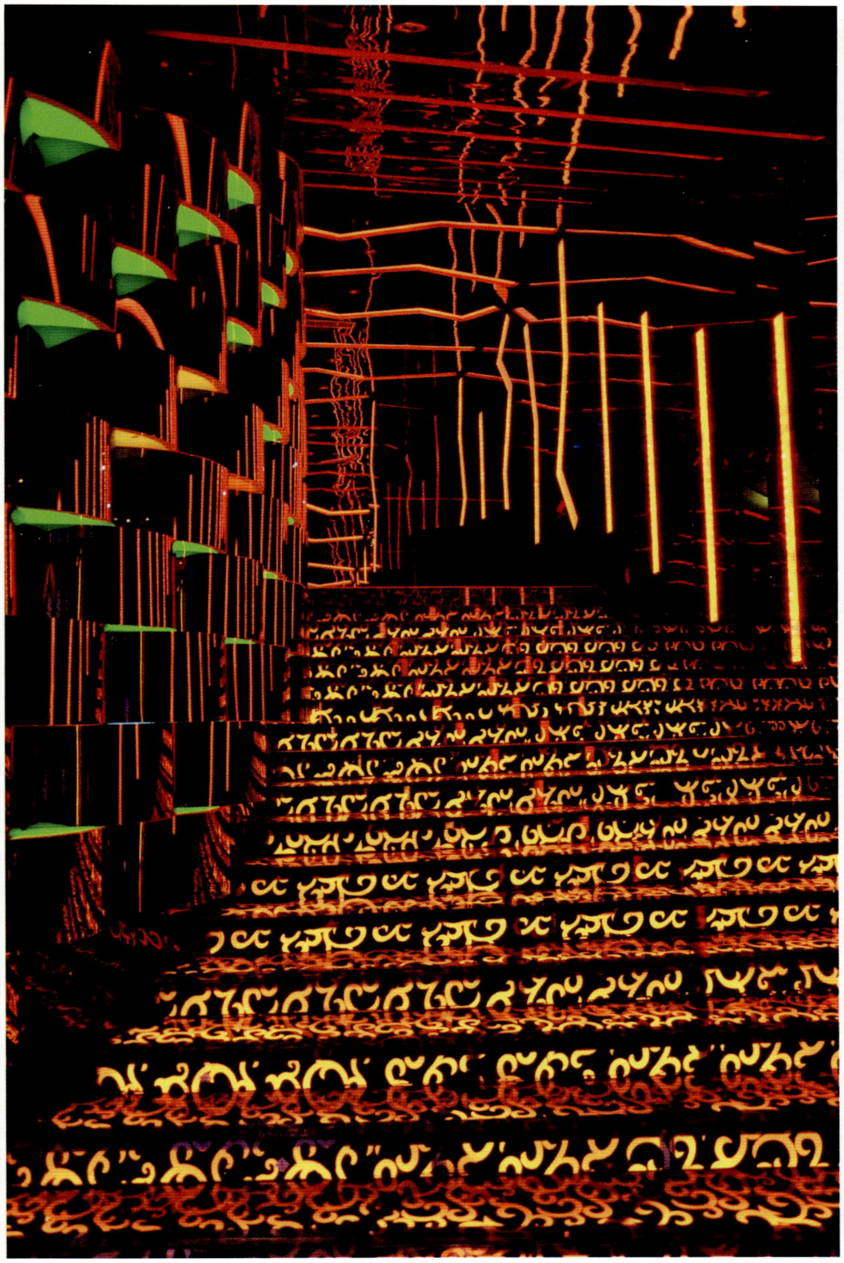

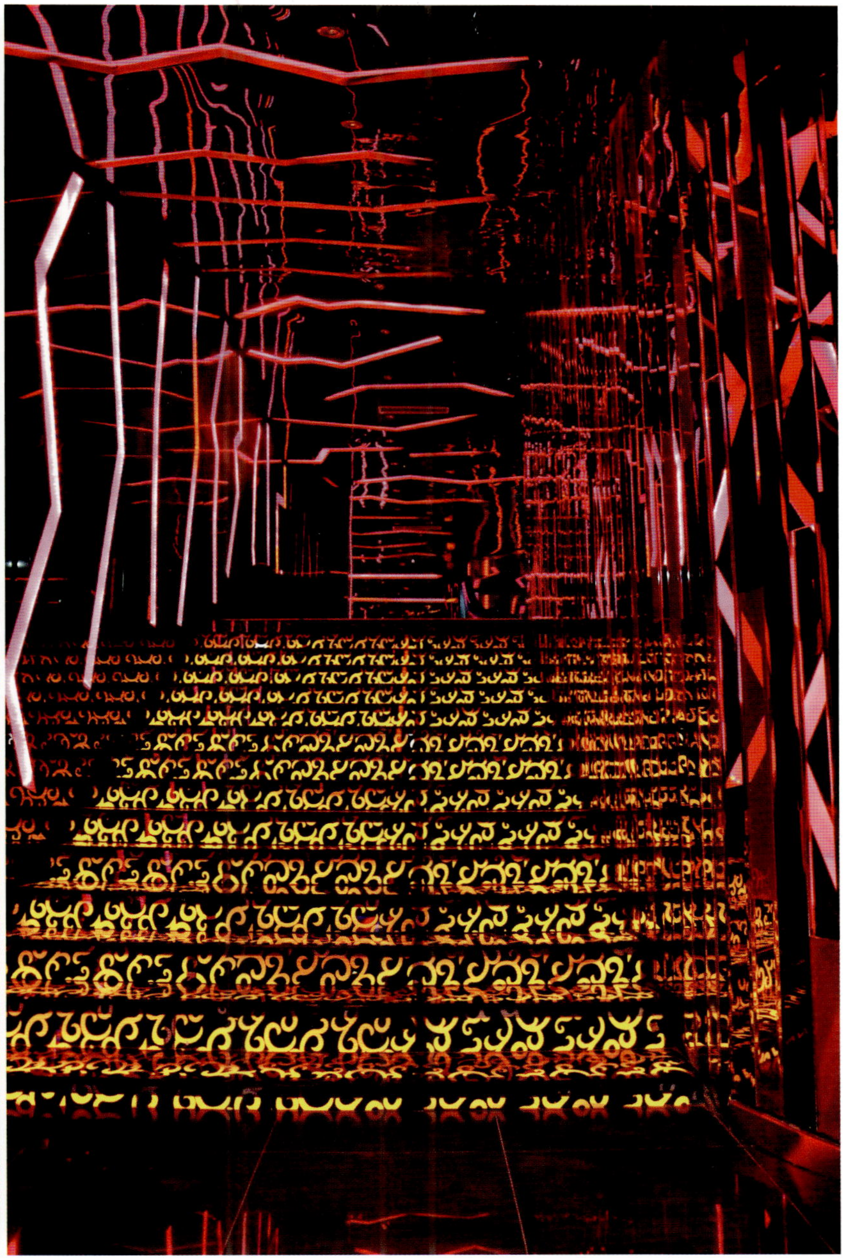

风格对娱乐场所而言是一种布局，也是一种品位的表现。正所谓设计决定成败，风格造就未来。本案设计敏锐地把握了市场的需求点，在室内设计上大量地选用有反射面的装饰材料，如深色玻化砖、黑玻璃、灰玻璃，配以坠帘、LED灯、图案壁纸、布艺沙发、发光玻璃茶几。通过设计师的巧妙构思，材料通过色彩、图形、灯光的作用组合在一起，在风格上得到统一与升华，在LED与玻璃镜面的映衬下，突显了整个东方娱乐城的流光溢彩、时尚浪漫之美，同时也展现出一种融合文化信息和娱乐功能的休闲空间之美。

深圳欢乐颂KTV

Shenzhen Happy Zone KTV

设计师：袁静、钟建福
设计公司：深圳市朗昇环境艺术设计有限公司
项目地点：中国深圳
项目面积：4000平方米
主要材料：大理石、地砖、涂料等

The KTV is named as Happy Zone, just as its name implies, it intends to bring happiness, joy to people all over the world. Positioned to target white collars in a bustling city, it's an entertainment place for busy white collars to obtain relaxation, leisure and pleasure. The image designs of exterior architecture, entrance and lobby feature rich, bold colors, varied looks, exotic accessories, making you experience a vivid image which is grand but not sumptuous, bright, healthy and free from inhibition. Such an image totally unveils the illusion usually found in nightclubs where people try to make them conspicuous but actually cover them up and people appear and disappear mysteriously. Internal functional areas include karaoke boxes, western food bar, supermarket, etc. Private rooms are designed to have over 10 types with different areas and grades to display varied and interesting audio and visual effects, satisfying diversified guests' needs. While the western food bar is designed with another mystic feel, where you can entertain your friends, relax, chat and slowly taste endless food and drinks.

魅惑

The designer boldly uses colors and design language to express exotic color elements .

运用大胆的色彩和设计语言，表现异域色彩元素，带给大众别样的心灵感受。

Heineken

欢乐颂 KTV
HAPPY

本KTV命名为欢乐颂，顾名思义，是要给大众带来欢乐、愉快的享受，它定位于繁华都市的白领，是忙碌的白领们在工作之余获得放松、休闲、快乐的娱乐场所。本案从建筑外观、入口及大堂处等形象设计上，色彩丰富大胆，造型缤纷变化，配饰充满异域特色，使人体会到的是豪而不奢、光明健康、淋漓畅快的鲜明形象，使夜场完全揭去那种欲彰还遮、神出鬼没的虚幻面纱。在内部功能设计上，主要为包房、西餐吧、超市等空间。包房设计十种以上不同面积与层次，以表现出丰富多彩的视听效果，从而满足不同类型客户的需求。西餐吧的设计，则是充满着另一种神秘情调，你可以在此呼朋唤友，休闲聊天，同时可以细细品尝无数的美酒与佳肴。

哈啰客艺术KTV

Hello Guest Art KTV

设计师：辛军
设计公司：深圳市辛视设计有限公司
项目地点：中国深圳
项目面积：1500平方米
主要材料：玻璃、地砖、乳胶漆等

Hello Guest Art KTV is located at Shenzhen, the world-famous Dafen Oil Painting Village, the existence of Hello Guest Art KTV breaks the traditional space arrangement, and it makes the entire room appear clear and clean. The relationship among rooms is more direct, and different rooms are linked together which are convenient for people to search and select.

The design of the lobby appears more contracted, and such design originates from the designer's understanding of colors. The modern sense and artistic sense are combined together which make the lobby appear more beautiful. People could see the colorful mosaic on the two sides of the wall at the passageway and the oil painting with various themes. The private rooms also follow such artistic flavor, some of the private rooms are based on contracted style, some of the private rooms are based on Chinese painting and some of the private rooms are based on abstract art.

魅惑

The oil painting elements express it's a space which combines artistic and modern design style.

油画元素表现一个艺术与现代相融合的空间，体现出强烈的时尚信息。

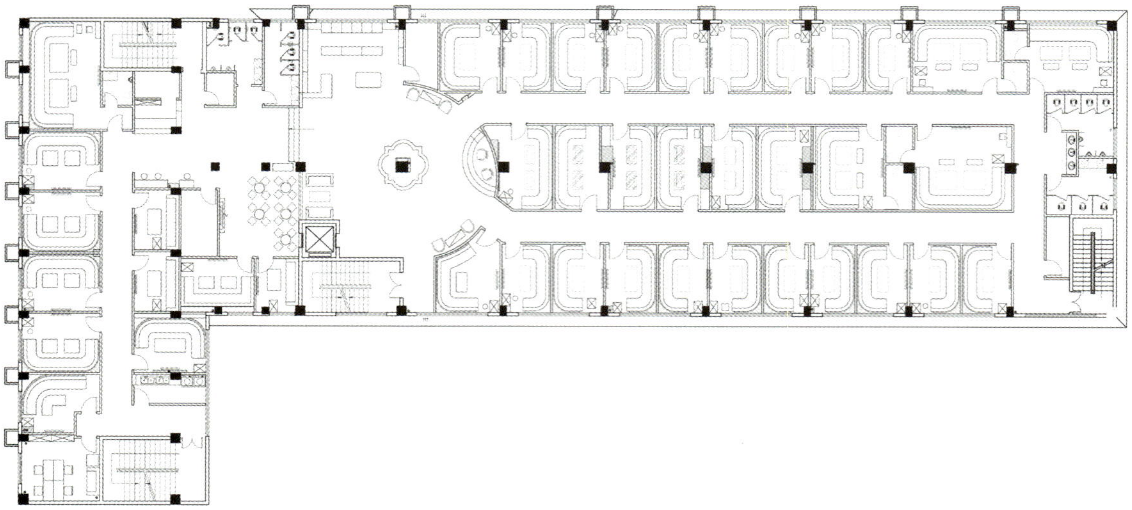

哈啰客艺术KTV位于中国深圳，地处世界闻名的大芬油画村，它的存在奠定了KTV的特殊文化背景。哈啰客艺术KTV的整体空间格局在传统KTV的基础上有所突破，实际上就是让整个空间格局更加一目了然。房间与房间之间的联系也更直接，不同的房间类型在一起，方便了顾客的寻找和选择。

大堂的设计更加素雅，这源于设计师对色彩搭配的理解，整体所产生的效果是非常明显的。一种现代感，一种艺术感，两者融合在一起，使得大堂更加深入人心。人们可以看到走道两边墙体上的斑斓的马赛克以及各种主题油画。包间里同样沿袭了这种艺术的情调，有的以简约的风格呈现，有的在中式绘画的基础上打造，有的以抽象艺术为基础。

A88

B20

7号公馆
Residence No.7

设计师：辛军
设计公司：深圳市辛视设计有限公司
项目地点：中国西昌
项目面积：约4000平方米

The concept of this case combines traditional art and modern element, making people could have the mental consolation while enjoying the material civilization. The appearance of Residence No.7 is simple and solid which shows the magnificent, generous and unique quality. The design of lobby continues the “magnificent” atmosphere, the sophisticated ceiling design uses golden foil, golden mosaic to decorate, and under the resplendent and magnificent lighting and glorious crystal droplight, the space shows us such luxurious, noble quality.

魅惑

Based on traditional elements and fashionable elements, Residence No.7 could afford the spiritual comfort to the people.

以传统元素结合现代时尚，让人在空间中得到精神的慰藉。

本案的设计理念将传统的艺术融合现代元素，使人们在享受物质文明的同时得到精神的藉慰。7号公馆外观造型简约、敦实，展现出了其庄重、大气的不凡气质。大厅继续延续“宏伟”的氛围，高深的天花，用金箔、金色马赛克拼花装饰，在灯光的映射下呈现出金碧辉煌的视觉效果，加上绚丽水晶吊灯的衬托，极具皇家贵族的奢华。造型水晶柱呈现宛若“疑似银河落九天”的意境，在变色的LED灯的照射下，仿佛不断上演着华丽变身的魔术。

V23

SEVEN

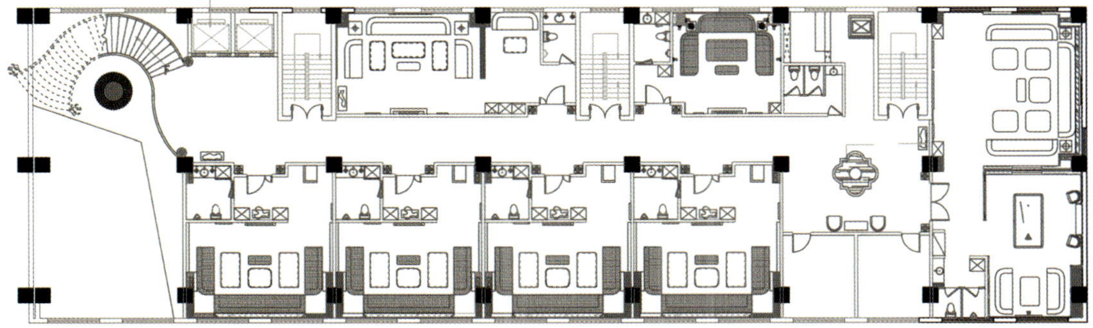

太子湾清音楼会所

Qingyin Club at Taizi Bay

设计师：刘枫华
设计公司：中国美术学院风景建筑设计研究院
项目地点：中国杭州
项目面积：1200平方米
主要材料：墙纸、铜雕、木地板、汽车漆、马赛克、饰面板等

The entrance hall is designed to combine both Chinese style and western style, emitting a strong neo-classical noble air. Bronze lattices divide the original windows in a natural way, serving the purpose of hollowed-out look, privacy and introducing outside views. Black serpeggiante marble floor brings out an imposing, dignified ambience. Cleanly shaped roof matched with gorgeous lighting fixtures also gives guests an innovative feel. Simple, generous entrance hall won't keep guests there long as they will directly enter the lounge area, where tea and refreshments are provided. The design method for this area is different from entrance hall. Here guests can chat or do business comfortably with a wide open vision. Windows, the intersection of indoor and outdoor, enable beautiful views outside to blend ones inside to let guests enjoy them all. Next is a long corridor and half-enclosed courtyard with tree. Where the courtyard meets inside space is the barrier of lattices, protecting the interior from the influence of changing seasons or wind, rain as much as possible. The compartment made of glass and lattices ensures an open, intimate environment, thoroughly showing a pleasant form.

魅惑

The designer focuses on the space function, and it has more humanistic feeling when the design extends its space.

注重空间功能的表现，更多的人性关怀
随着空间的延伸逐步体现出来。

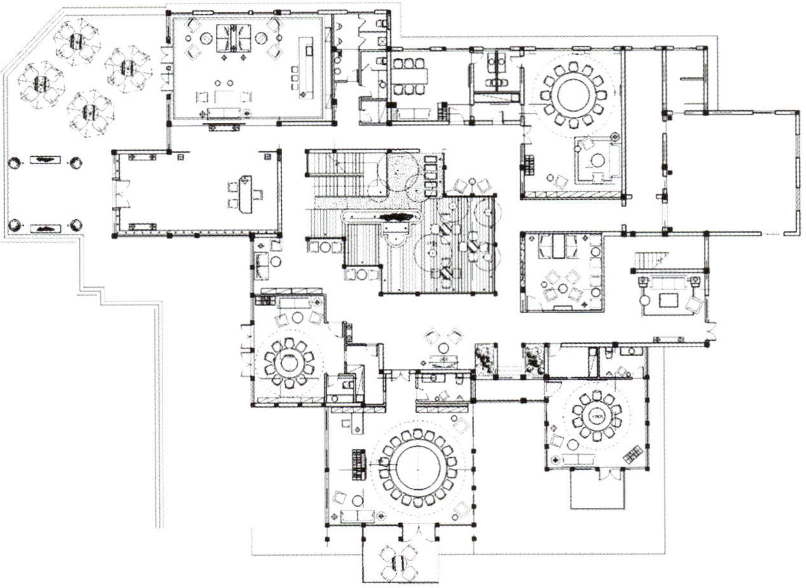

一层

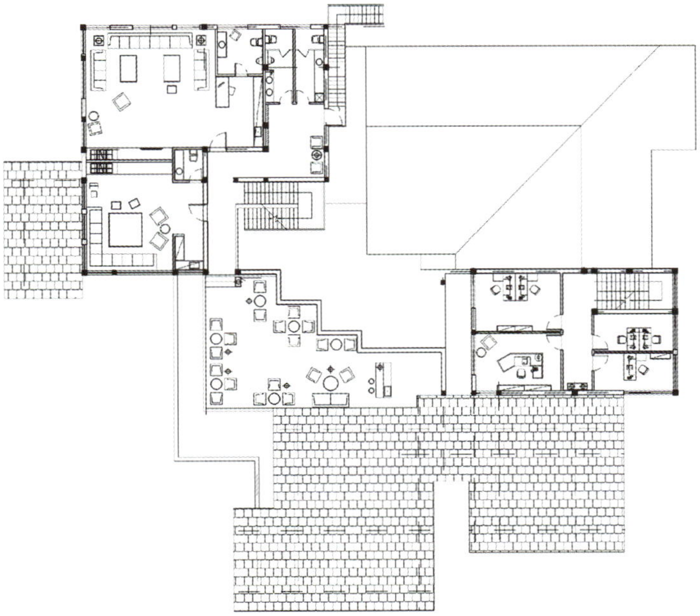

二层

门厅设计中西结合，有着很强的新古典贵族气息。铜制花格隔断，将原有窗户自然地分隔开来，花格既有透空造型，可看到室外景观，又有私密性。地面黑色木纹石，整体大气、稳重。顶面造型简洁，所选用灯具个性鲜明，目的也是让客人产生新颖的感觉。门厅简洁大方，客人不会在此停留很久，将直接进入休息区。休息区为客人提供茶水点心等，这里设计手法区别于门厅，客人可以在这里很舒服地聊天以及洽淡生意，视线也开阔。窗户是室内室外的交界点，室外的美景融入室内，让来客享受于其中。穿过长长的走廊，进入半封闭的庭院。院中有树。庭院与室内的连接处，用花格隔断。在四季变化、刮风下雨等情况下，室内环境尽量不受影响。采用玻璃隔断加花格造型，既通透又私密，使人尽享美丽的景致。

重庆凯撒王宫国际俱乐部

Chongqing Caesar Palace International Club

设计师：吴恙
设计公司：深圳市吴恙室内装饰设计有限公司
项目地点：中国重庆
项目面积：1500平方米
主要材料：玻璃、大理石、金属、布艺、亚克力等

The designer of Chongqing Caesar Palace International Club expresses the popular, fashionable and abstract design through modern decorative materials. From the pictures, the modern decorative materials are throughout the double-deck rhombus. The image inspiration of industrial sense of "circuit board" and the adornment huge flower come from the model's cloth fabric of the top-level fashion show and the experience accumulation of life. As for the materials, the designer catches the major feature of interior, using bright or rigorous, tough or soft, luminous or nonluminous materials to make contrast and express its modern design. As for the lighting, the designer uses transparent blue light and mysterious purple light to express his design idea, and decorated with LED and EL changeable light.

魅惑

The designer uses abstract and popular design language to express the space and uses modern materials to decorate the space.

用流行而抽象的语言表达空间，用现代的材料装饰空间。

重庆凯撒王宫国际俱乐部设计师通过现代的装饰材料来表现流行、时尚、抽象的艺术语言。从图形语言上，设计师贯穿整场使用双层边菱形、具有工业感的“电路板” 形、具有装饰感的大花形，这些图形灵感大多来源于世界顶级时装发布会模特的衣料或生活的积累。在材料表现上，设计师抓住了室内装饰材料的主要特点，使用光亮或粗糙，坚硬或柔软，不发光或发光的材料通过对比和构成的手法来表现前面所描述的时尚语言。在灯光设置上，以“深远”、“透明”的蓝光和“暧昧”、“神秘”的紫光来表达，配以局部的LED、EL可变色灯光。

麦乐迪量贩式KTV

Melody KTV

设 计 师：李景光
设计公司：上海景光室内空间设计创意坊
项目地点：中国山东
项目面积：2600平方米
主要材料：花纹深灰地砖、PVC圆管喷漆、混油漆、黑玻璃、灰镜、金属马赛克等

Commissioned by a nightclub in Shandong province, the designer finished the design for Melody KTV. These are real-scene pictures of the second phase engineering project, which is located on the second floor. The project covers total 2600 square meters with small pub in private rooms, for which the design is mainly focused on the hallway, corridors and the environment of reception desk on the second floor, the overall height being 2.3 meters. Laying an emphasis on linear sense, especially the soft sense of lines, the design, coordinated with lightings, produces a pure and beautiful feeling of the space. Constantly varying lightings bring rich visual senses and people are fully aroused into high spirits along with the changing lights.

魅惑

The expression of line combines the using of lighting, which expresses the pure and elegant sensation.

线条结合灯光的运用表达了单纯和优雅的感觉。

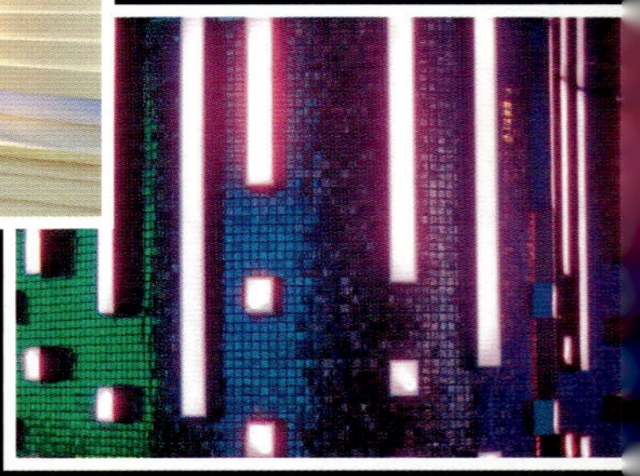

此项目为设计师在山东省设计的夜场，即麦乐迪量贩式KTV庭场设计。实景图片为此项目的二期工程，位于二楼，总面积2600平方米，包间另有小型轻吧，重点设计了过厅通道走廊及二楼服务台的环境，整体高度为2.3米。设计注重线条感，特别是柔软的线条感的塑造，配合灯光的运用，为空间营造出单纯和优雅的感觉。灯光的不断变化给空间带来丰富的视觉感受，人随光动，情绪被很好地调动起来。

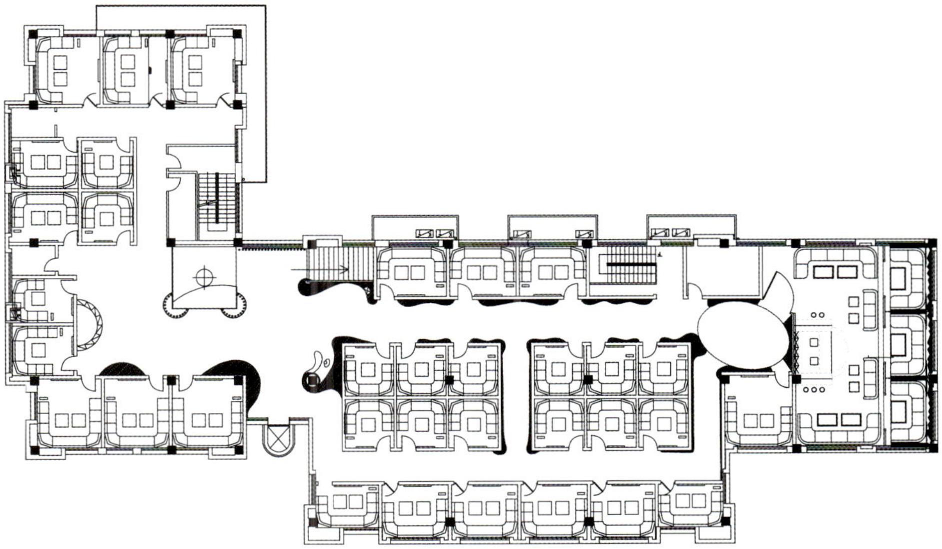

金歌KTV娱乐休闲中心

Jinge KTV Entertainment Center

设计师：程英杰
设计公司：上海九轩室内设计咨询有限公司
项目面积：2000平方米
主要材料：黑镜、镜面马赛克、钛金板、烤漆玻璃、镜面夹丝玻璃、LED灯、墙纸、地砖等

Modern, fashionable, gorgeous are the theme of this case. The designer pursues the simple design, and through some reflective materials to enrich the entire space. The changeable and dynamic LEDs are displayed throughout the interior space, together with background music, adding the recreational air to the space. The house number of the box is presented trough spotlight projecting the light on the floor, which appears fashionable and chic. This is a reconstructive case and the public space is very low in floor height. The designer combines large scale black mirror and LED with frosted glass, making the lobby and hallway more spacious and variegated.

魅惑

The abundant colors and formations highlight the excellent quality of the space , as well as the elegant detailed design.

丰富的色彩和造型，以及精致的细节打磨，突显出良好的空间品质。

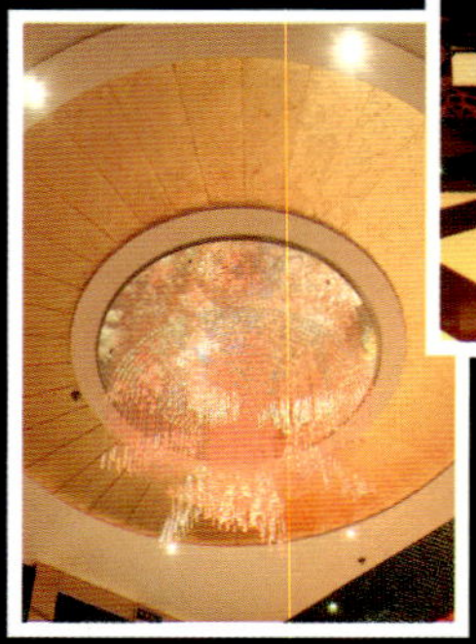

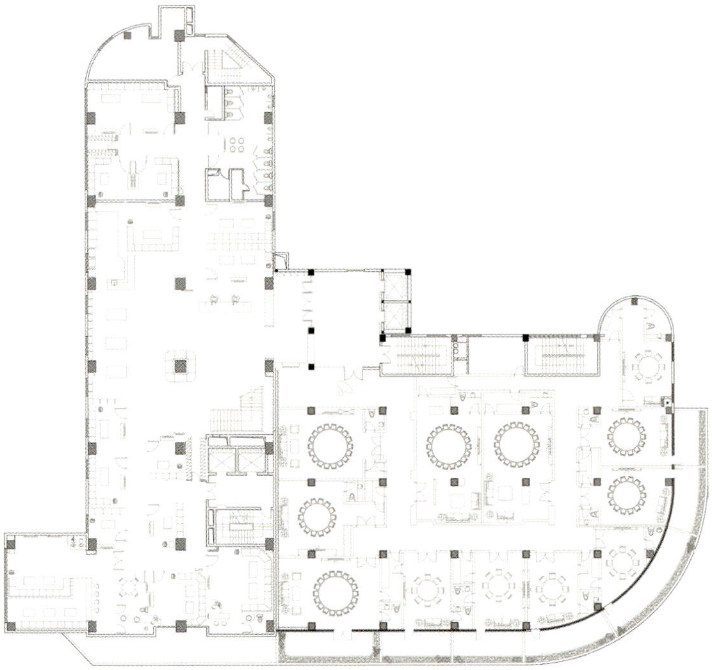

一层

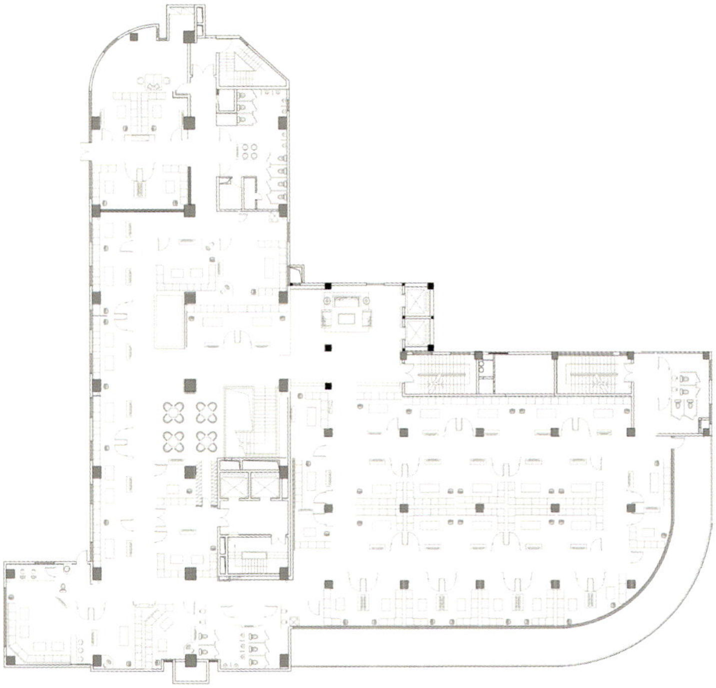

二层

现代、时尚、绚丽是本案的主题。力求用最简洁的造型，通过一些反光材质的应用来丰富整个空间。动感、变幻的LED灯贯穿整个室内空间，配合背景音乐，更加增强了空间的娱乐性。包厢的门牌号是用聚光射灯打在地板上的，新颖、时尚。本案为旧楼改造，公共空间层高很低。设计师将大面积黑镜和磨砂玻璃材质的隐藏式LED灯结合起来使用，使得大厅与走廊更加开阔且富于变化。

宁波永利国际KTV

Ningbo Yongli International KTV

设计师：许国维
设计公司：宁波东道建筑装饰有限公司
项目地点：中国宁波
项目面积：2800平方米
主要材料：玫瑰金不锈钢、灰镜、黑镜、啡网纹、爵士白、珊瑚红、成品UV板、镜面马赛克等

The outstanding feature of this project is to pursue the varying beauty and the changes in styling, color and lighting are all enough to arouse people's interest in the space. Full of strong classical air, the wall and floor in vintage colors bring people back to the ball room of 19th century. The changes in styling are reflected from everywhere your vision might reach, for instance, when walking along the corridors, the looks along both sides of corridors are transformed so cleverly and delicately just like changing scenes in a movie. The rendering of colors leads to an extremely sumptuous and trendy KTV. Meanwhile, the project is also concerned about guiding customers psychologically. Different wall windows are installed at some originally unnoticeable areas to display works of art, beers, drinks, etc. Here people will find experiences beyond traditional KTV.

魅惑

The changeable feeling could make customers have fresh feeling, and the subtly using of colors and lighting abandon the traditional restriction.

变化之美让空间时刻带给顾客新鲜感，巧妙的色彩和灯光运用，摆脱了传统的束缚。

HELAN
MOUNTAIN

贺兰山
V1006

TCL

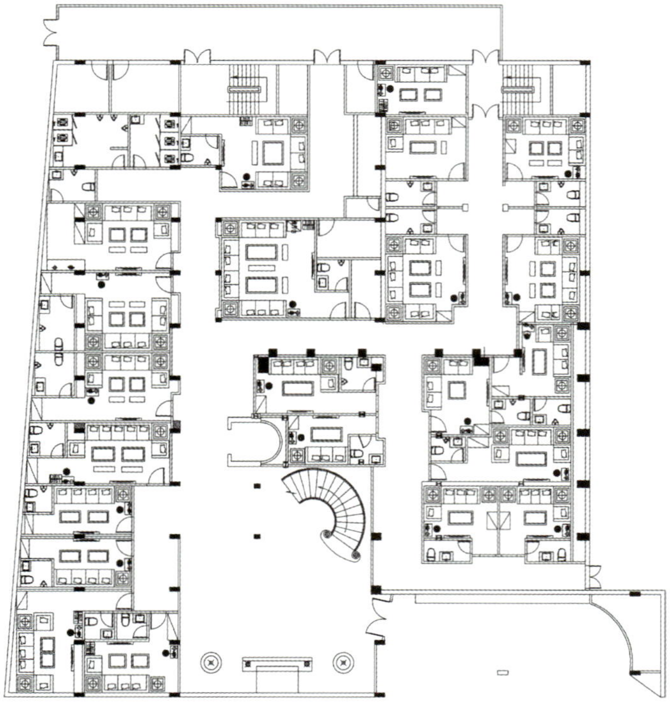

一层

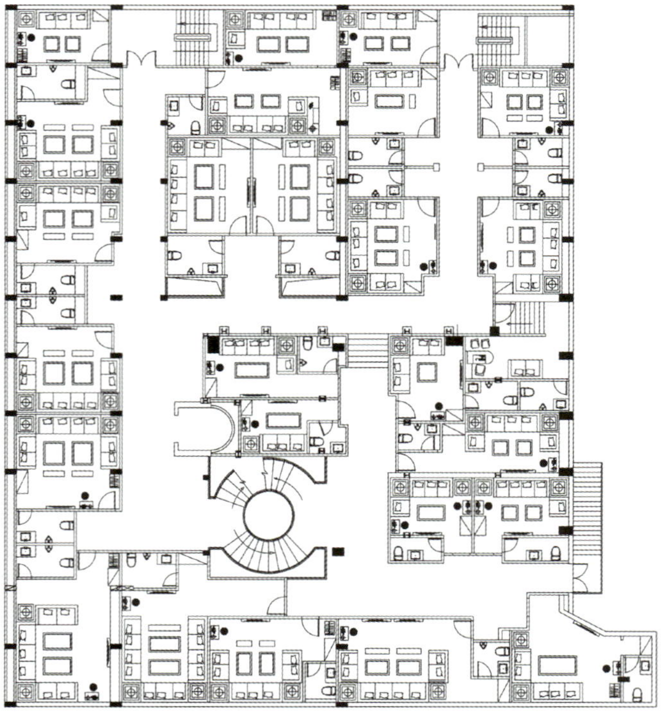

二层

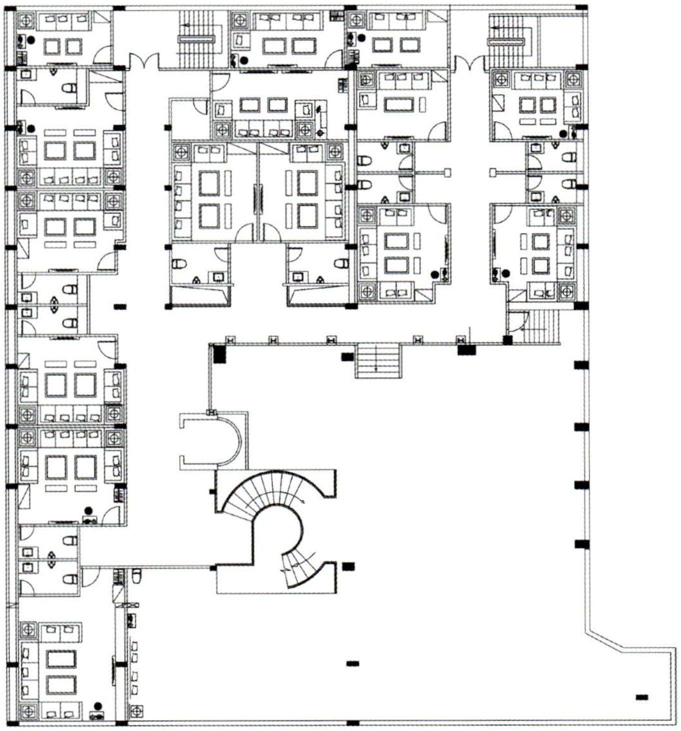

三层

追求变化美是本案的最大特色，造型上的变化，色彩上的变化，光线上的变化，都足以引起人们对空间的兴趣。复古色彩的墙面以及地板装饰，将人们带入了19世纪的舞会现场。这里洋溢着一种浓郁的古典气息。本案追求的造型上的变化体现在视线可以触及的每个地方，当人们从走道上经过的时候，走道两边的造型转换犹如电影场景一般精致且巧妙。色彩上的渲染，打造出了KTV的极致奢华感与时尚感。同时，本案关注对顾客心理的引导。在一些原本不起眼的区域，设置了不同的壁窗，里面陈设着艺术品以及啤酒饮料等，让人们在这里可以找到超脱传统KTV的感受。

紫金城夜总会

Zijincheng Nightclub

设 计 师：张振斌
参与人员：刘艳娥、王林、谢宇哲、扬帆
设计公司：华夏建筑装饰工程有限公司
项目地点：中国烟台
项目面积：3500平方米
主要材料：黑金砂理石、印度红理石、木雕、玻璃钢、透光石、仿古砖、羊皮、钢化玻璃、艺术肌理漆等

This building is the royal residences of Emperor QinShiHuang built in order to commemorate his east tour. Interior design follows the traditional building regulations , to embody the historic background and also satisfy the function of evening get-together. With the companion of nature seneries you will find on the left side history stories are lifelike , while on the right side a red sheepskin stamped with poetry fully embodies Chinese culture. Around the whole hall, you can feel the vibration, thus conclude that nature and palaces are grand and such luxurious palaces embody the honor of QinShiHuang. Stop at the doorway you will see the hallways with 30 lanterns, deliberately revealing a palace of ever-roaring grandeur and also the spirit of design.

Walking through the lobby, step into "space-in-time bar" you'll feel the historic atmosphere of east tour on one side ,and the other side is modern music bar , you can walk through both sides and also the time zone , with this constant everlasting music lingering the total five thousands years.

魅惑

The designer fully uses the natural scene and makes people can feel the power of nature world.

充分利用自然景致，让人感受到自然的力量。

本案建筑是为了纪念秦始皇东巡而建设的一座“行宫”。室内设计遵循外建而围绕中国历史长卷展开思绪，既体现历史底蕴，又要满足夜场功能，充分利用自然场景，左侧历史演义，右侧中国红羊皮印着诗经，充分展现中国历史的文化。环绕整个夜场，给人以震撼的效果，感到自然之大、宫殿之雄伟、选山之景之险峻。彰显始皇疆海无界、皇帝之尊，充分体现宫殿之豪华。驻足于门厅，展现于眼前的为30盏灯笼的玄关，刻意彰显宫殿的磅礴气势，诠释设计的灵魂。

穿越大堂登上“穿越时空静吧”，一方是始皇东巡，另一方是现代音乐酒吧，任现代穿行于古典，让音乐横亘古今，缭绕五千年之时空。

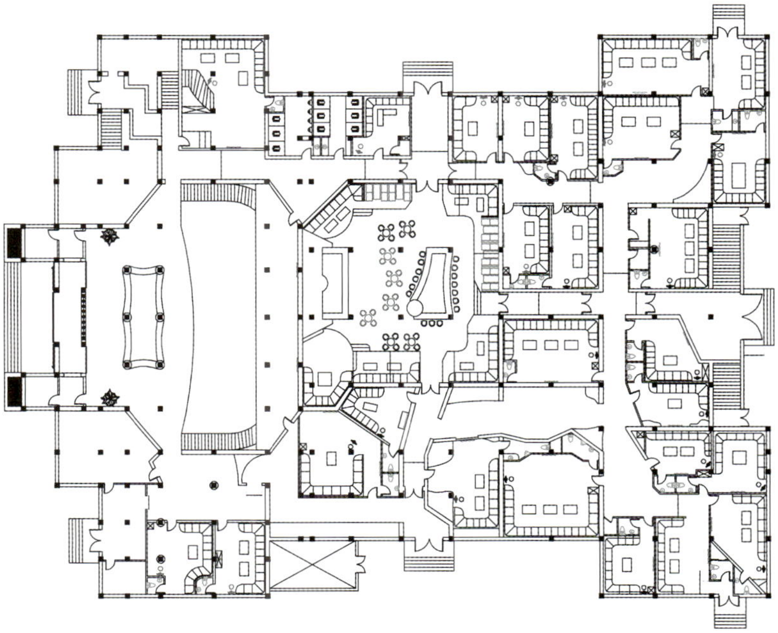

Fascinating Bar

Slow rock, mellow wine
Dazzling light, comfortable chair

Bar has been a part of urbanites' life long time ago
Maybe you could enjoy yourself with your friends
Maybe you could have a rendezvous with your love
Maybe you're alone
Every person could get what he wants here

It's not only about the wine, but also about the atmosphere
A place where you could relax yourself
Here you could forget your trouble and time
You could also leave the busy city life aside
Here you could enjoy fair-sounding music and mellow wine

This is not an art palace
But you could see the artistic signs here
Such as Bar of Santiago
With a kind of European neoclassical flavor
And fashionable and luxurious flavor as well

The memory of this place is not all about city life
Such as Greenhouse Nightclub, people inside could feel like living in the natural world

Such as Suzhou Royal TT Bar
It merges the cultural feature of Suzhou
Based on red and yellow, decorated with blue and white
The design creates the warm and cozy feeling

Different bars have different features
Different bars are with different themes
Maybe it' about music
Maybe it's about reading

From designer to guest, from singer to bartender
Bar provides the suitable wine for each person

The design of the designer is about bar design
It is also about the place for modern people who enjoy the nightlife

魅惑·酒吧

慢摇，芳醇的美酒
迷离的灯光，舒适的桌椅

酒吧，早已成为都市人日常生活中的一部分
也许是和朋友在一起欢乐
也许是和另一位在此约会
也许只是一个人在此独自消遣
酒吧包容着城市里的每个人

酒吧提供给人的不只有酒，还有一种氛围
一处可以释放个性的空间
这处空间，可以让你忘记烦恼、忘记时间
一处可以撇开都市忙碌生活的乐园
这处乐园，可以让你尽享动听的音乐和芳醇的美酒

这里不是艺术的殿堂
但是可以看到艺术的影子
如同圣地亚哥酒吧
带有一点欧式新古典
却又夹着一点时尚与奢华

这里不是完全的都市记忆
正如绿色酒吧，仿佛让人身处大自然

苏州皇家TT酒吧
融合苏州这座城市的文化特质
以红黄为主，点缀以蓝白
营造温暖与舒适

不同的酒吧有着不同的特点
不同的酒吧有着不同的主题
也许是音乐酒吧
也许是阅读酒吧

从设计师到顾客，从歌手到调酒师
酒吧酿出了适合每个人的一杯酒

设计师们设计的是酒吧
也是现代人夜生活的栖居地

北210

210 North

设 计 师：Charles Doell; Gui Bez
设计公司：大人物设计
项目地点：美国里诺
项目面积：743平方米
项目摄影：Jeff Dow

210 North provided a physical space that was a perfect foil for designer's interest in creating spaces that are not only glamorous, in a traditional sense, but also modern, irreverent and exuberant.

210 North had a previous life as a casino, so designer had inherited a large dark staircase entrance that we transformed into a stunning entry with chain curtain and saturated colored lighting for the two sidewalls. The entrance was very glamorous but the psychedelic aspect took the viewer to a more oblique place. In the main spaces the designer used bouquets of the same LED shade lamps to help define different areas in the club, each area with its own special feature. Most notable is the huge suspended glass shard chandelier by Eva Menz. This chandelier is definitely glamorous and technically modern.

魅惑

The exaggerate vision tensility is expressed by the attractive colors, as well as the modern and traditional elements.

以俘获人心的色彩表现出强烈的视觉张力，吸引人眼球的还有现代元素与传统元素的结合。

N
210

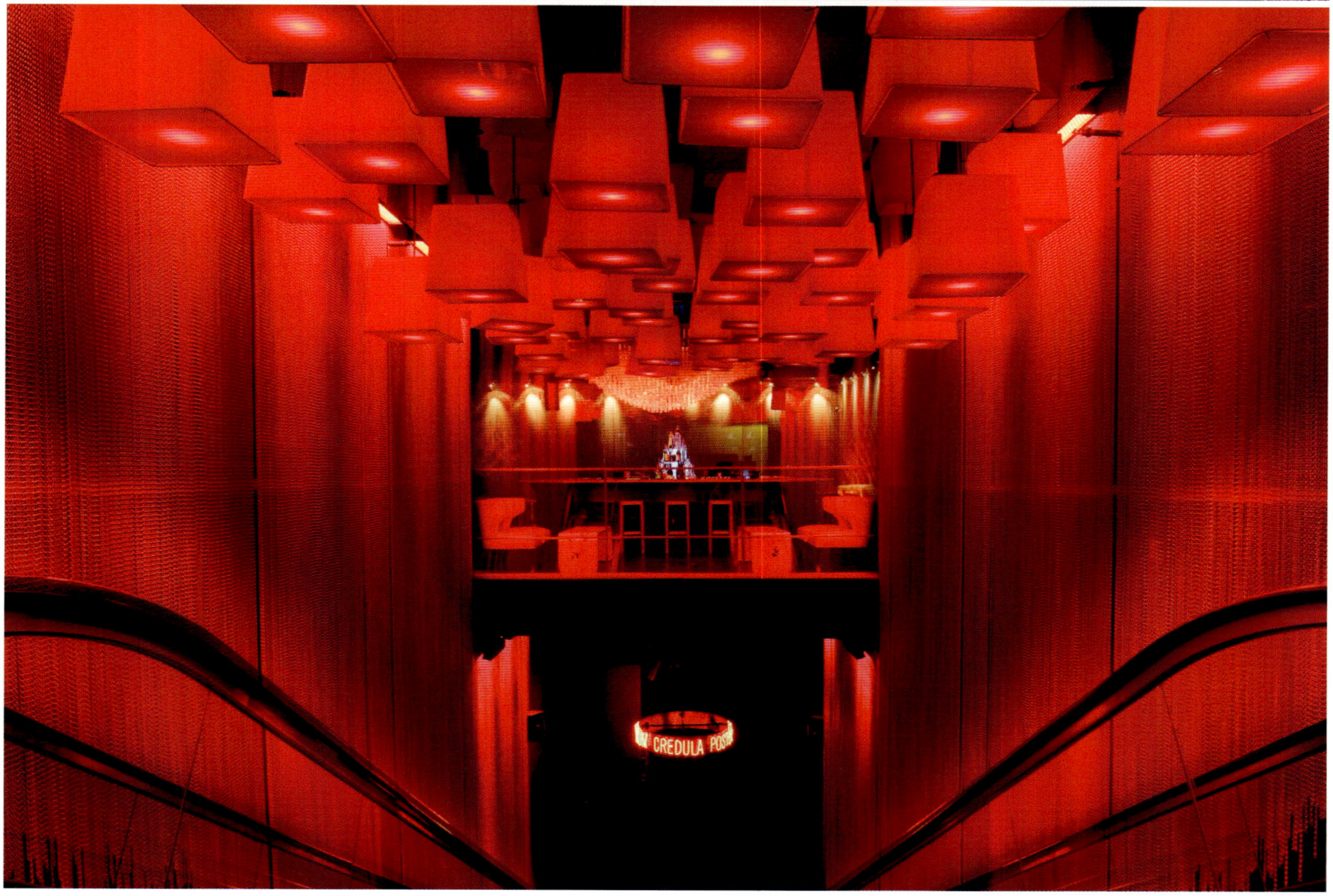
CREDULA

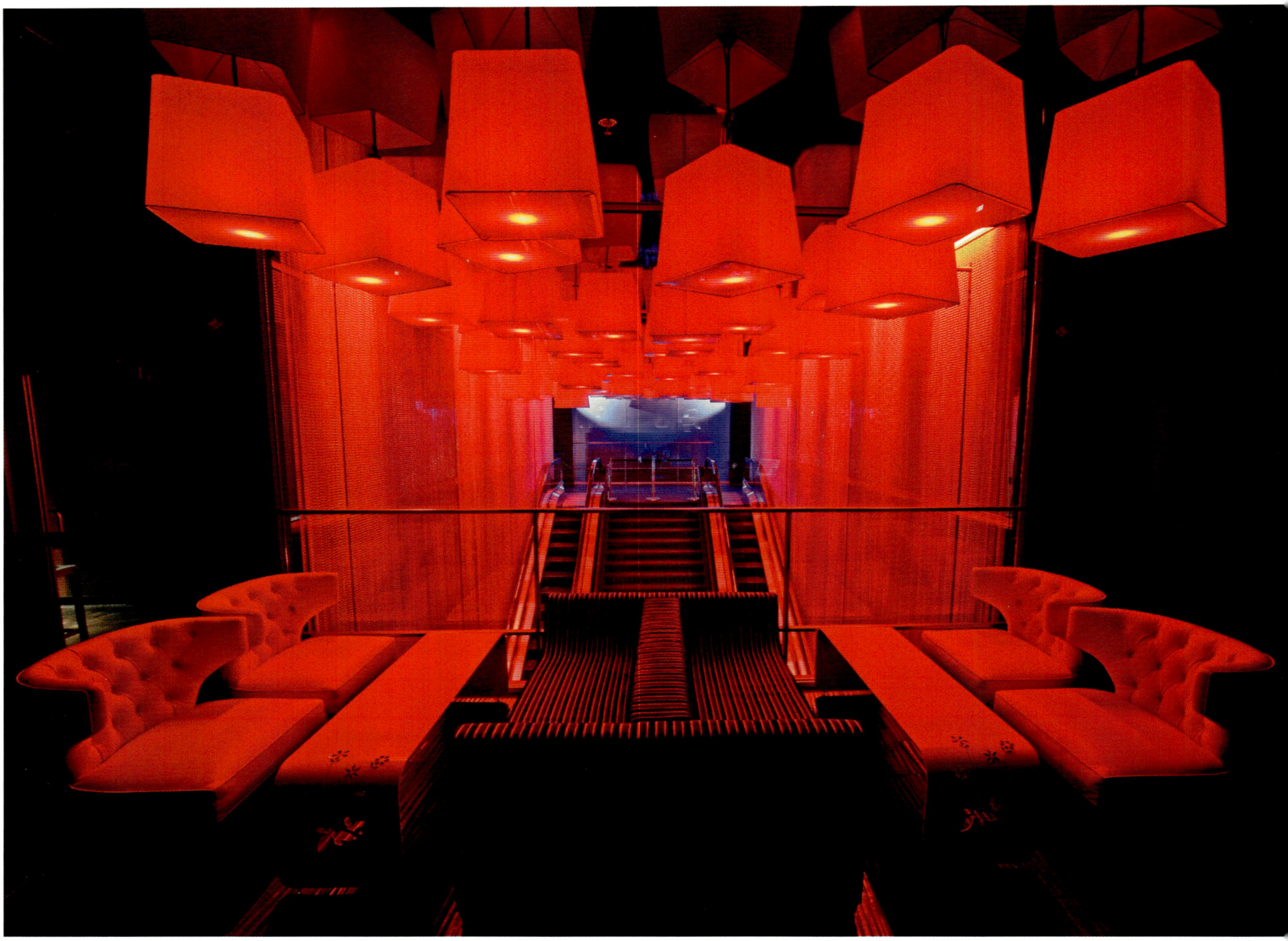

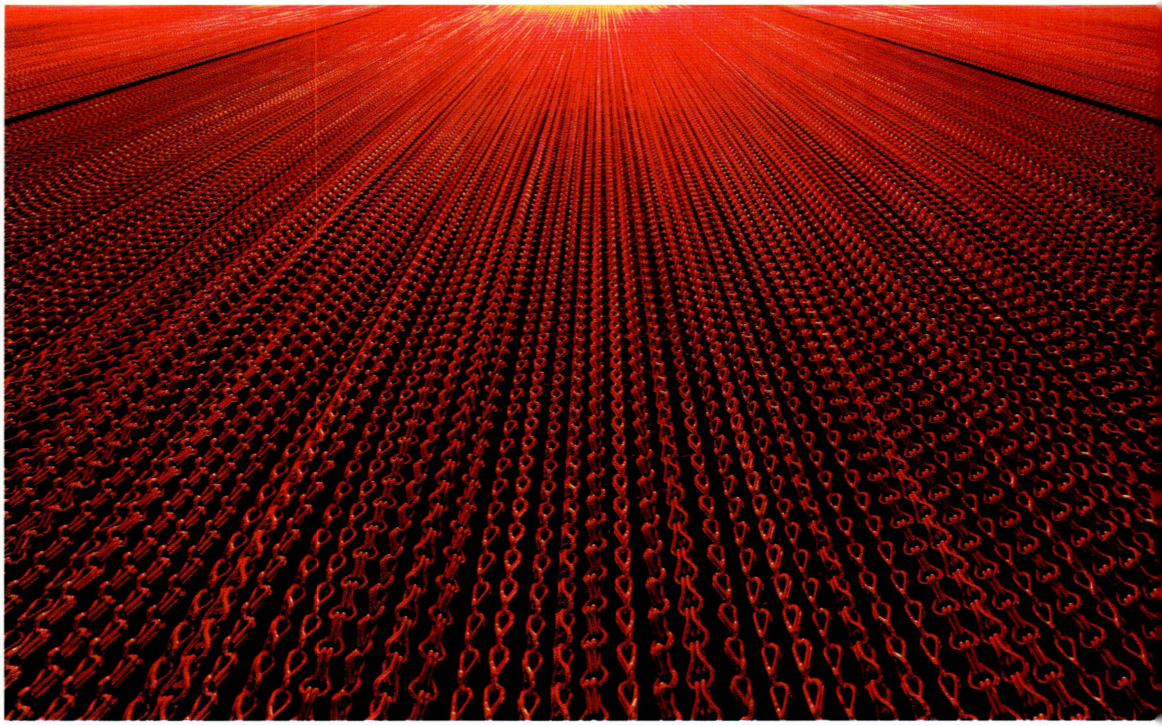

北210为设计师提供了一个充满情趣且星光闪耀的实体空间，在此空间里，现代元素与传统元素并存，设计元素丰富且不着边际。

北210以前是一个娱乐场，设计师用窗帘链把之前的大型黑色楼梯入口改造成一个极富魅力的入口，并在两边用绚丽的色彩装点墙面。入口处的设计引人入胜，充满迷幻色彩的造型把来客带入了一个倾斜的空间。在主空间里，设计师同样运用了LED来点缀整个会所的不同空间，每一个区域都有它独特的空间特点。最引人注目的要数由Eva Menz设计的巨大的悬置玻璃枝形吊灯，这个枝形吊灯非常绚丽且现代。

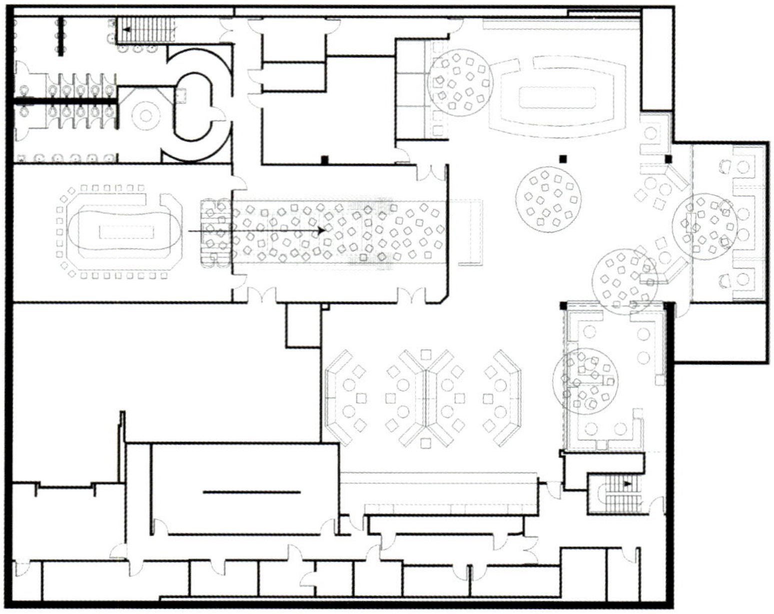

SET酒吧

SET

设计师：Francois Frossard
设计公司：FFD Inc
项目地点：美国拉斯维加斯
项目面积：930平方米

SET features a "euro-hip" go-go style interior. Francois understood the mistakes of his predecessors and drew on Miami's "modern" European style. Although SET is majestic, it is free of pretension and a satire of the buildings' former lives. François installed a working glass fireplace; each piece of furniture is covered in luxurious Pucci fabric and the Chihuly inspired chandeliers mock authority.

魅惑

SET features a "euro-hip" go-go style interior, and it also merges modern design style.

SET酒吧以“euro-hip”摇摆式室内设计风格为主要特征,并融进了现代风格。

SET酒吧以“euro-hip”摇摆式室内设计风格为主要特征，Francois清楚地认识到他的前辈在以前的案例中的失误，并且借鉴了迈阿密“现代”欧洲风格。虽然SET看起来宏伟、壮丽，但在之前却并不被人赞许。Francois还为这个空间装设了一个透明壁炉。空间中的每一处陈设都是由豪华的Pucci织物以及充满创意的Chihuly枝形吊灯点缀的。

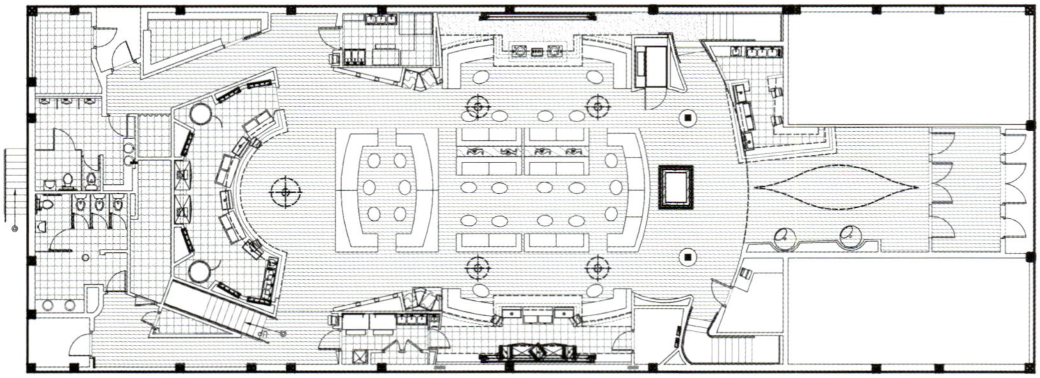

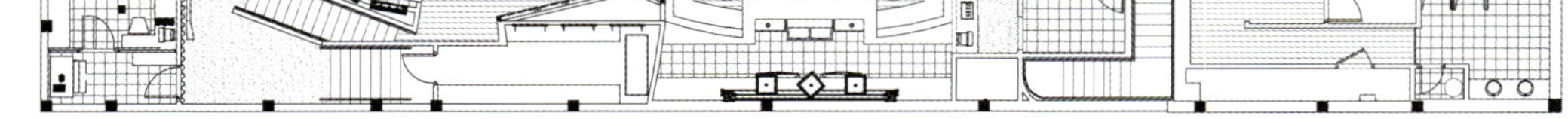

绿色酒吧

Greenhouse Bar

设计师：Antonio Di Oronzo
设计公司：bluarch architecture + interiors
项目地点：美国纽约
项目摄影：ADO

Greenhouse is a 4,000sf nightclub, lounge and event space built from recycled or recyclable materials, which is the first nightclub in the nation to receive certification via LEED_CI by the United States Green Buildings Council for its environmentally conscious construction and design. The designer decided to stay away from recreating a greenhouse, and opted to transpose the notion of landscape to an interior space. The design concept was to convey the dynamic richness of nature as a living system. The tables are a quirky presence in the overall scheme. They are tempered glass boxes containing boxwood covered wireframes shaped as different animals resting on a carpet of eco-friendly, artificial grass.

魅惑

The green elements can be seen everywhere, and the entire design focuses on natural feeling and green life.

绿色元素随处可见，注重表达自然感受，倡导绿色生活。

EXIT

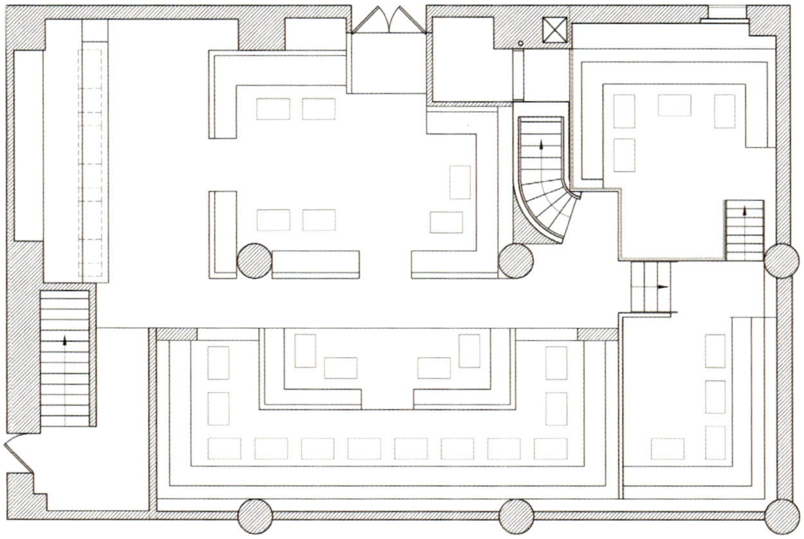

绿色酒吧是一个371平方米的夜店，所有的沙发软座、聚会空间的设计材质都是通过回收加工的。这个绿色酒吧是美国首例拥有此类设计的酒吧，并且荣获由美国绿色建筑委员会颁发的荣誉证书。设计师决定放弃单纯地重新塑造一个绿色酒吧形象的思路，而更愿意把自然景观带入室内。整个设计理念都是为了表达富有活力的自然生态系统这一概念。桌子都以怪异、新奇的造型出现在整个空间里。桌子是钢化玻璃制成的盒子，里面有由线框做成的植物雕塑。雕塑以不同的动物造型呈现，脚下便是环境友好型的人造草。

保龄球–台球俱乐部

Bowling Billiards Darts Club

设计公司：PARASITE工作室
项目面积：2000平方米

As a relaxing club, this space affords us the complete facilities, which could make people have the joyful feeling while doing some body movements. The space focuses on the relaxing manner, making people could find some relaxing feeling.

This contracted club is situated in an iconic building; inside the club, people could hardly know the variations of the outside world and the flowing time. The designer wants to give fresh space air to the guest through such space design. The design originates from blackbox design concept. The designer searched some design elements through blackbox design concept, and finally he decided to create an abstract space. The existent window becomes the information exchanges platform and exhibition screen.

魅惑

The designer uses the concept of "blackbox" which creates the fresh new space and thus distinguishes the interior and the exterior.

用“黑盒子”的理念，打造全新的空间，让空间和外部世界脱离开来。

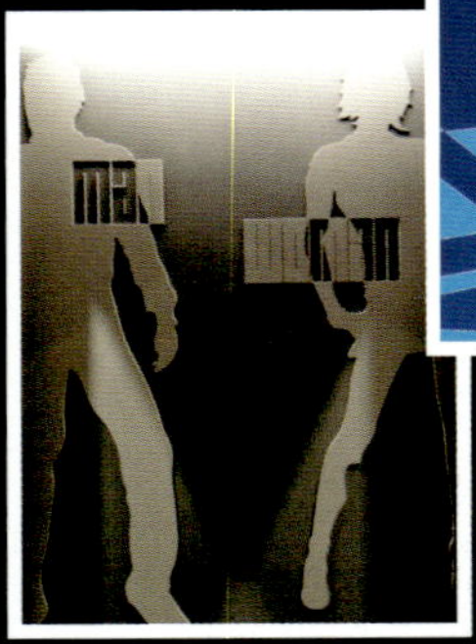

bar

作为一个休闲俱乐部，这里提供了完善的设施，供人们在进行身体运动的同时，也享受着精神上的愉悦。空间集中表现了一种休闲的方式，使得现代人在都市生活中更加容易找到一种放松的感觉。

这个简洁的俱乐部坐落在一个具有标志性意义的建筑体之中，在俱乐部的内部空间中，人们很难意识到外面世界的变化或者时间的流逝。设计师想通过这样的空间带给顾客一种全新的空间氛围。设计方案产生于黑盒子的理念，设计师从中寻找适合本案的一些元素，最终决定将空间打造成一个抽象的空间，现有的窗户成为一个信息交流和展示的平台。

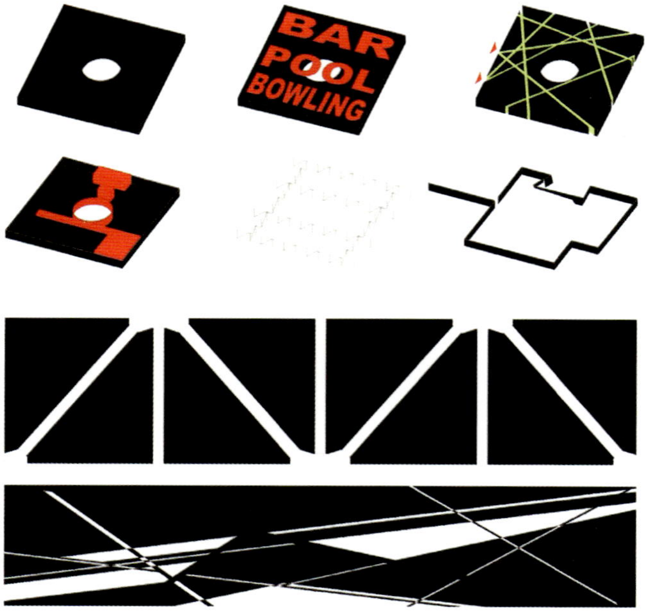
BAR
POOL
BOWLING

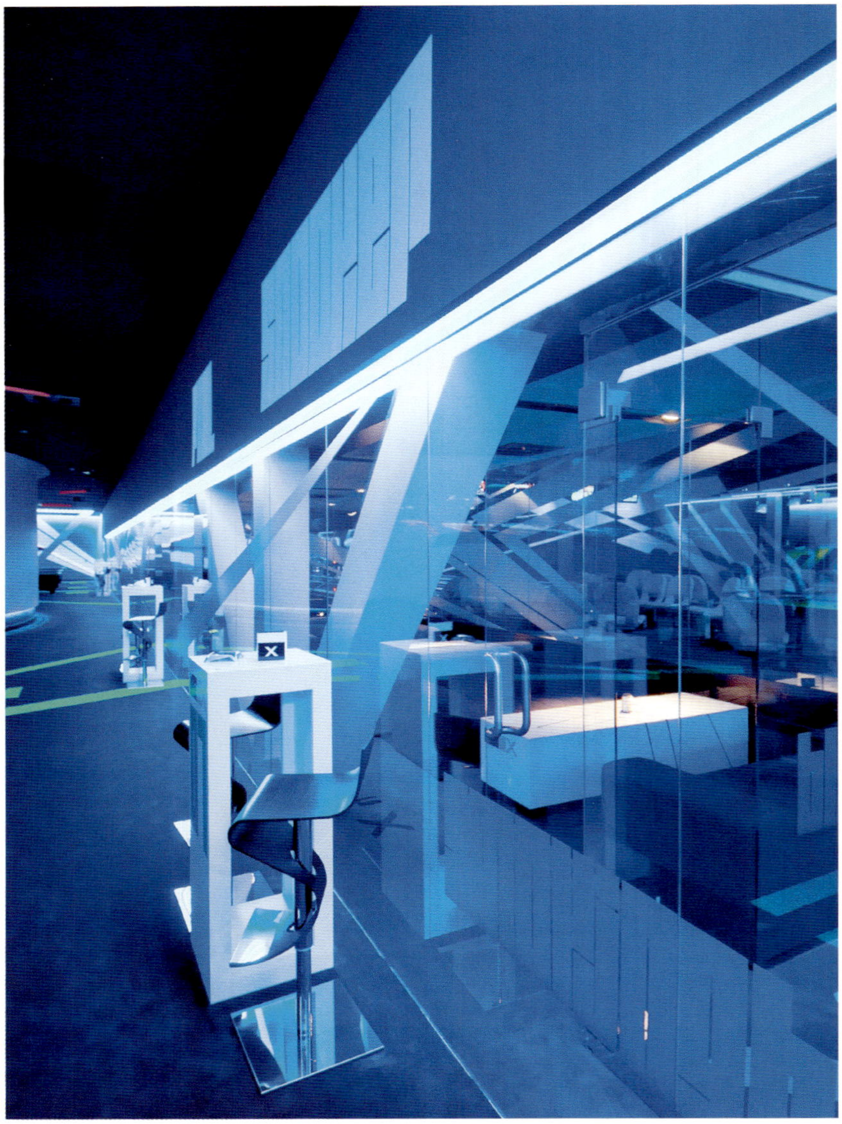

invention of
inspotter
1940s
Bowling

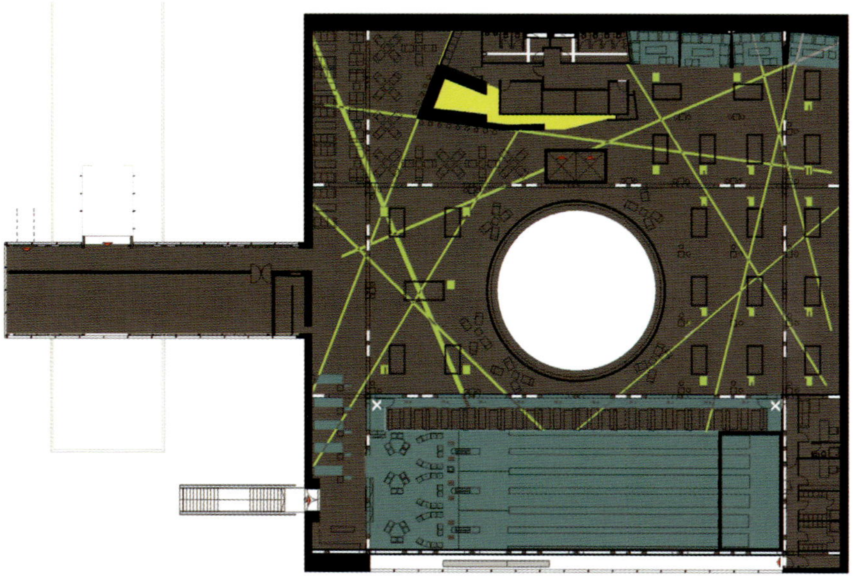

上海Drop酒吧

Shanghai Drop Bar

设计公司：OOBIQ Architects
项目地点：中国上海

At the beginning of the design the designer needed to define what was right in terms of design. Finally the designer found contemporary classic is the right definition of this place.

The building where Drop is located is one of the few buildings in downtown Shanghai with such original decoration on the ceiling and on the walls. It used to have its own character and soul. With the definition of "contemporary classic" OOBIQ Architects refers to an interior made of classic textures, traditional shapes and materials, but transformed and adapted to a Club to be located in such an international city as Shanghai. The whole club is presenting the same floor finishing all around the areas: A geometrical Moroccan tile became the motif of the precious marble floor. This warm decoration becomes another element that shows clearly the crossover of the interior design of Drop: Elements of different cultures and times connected together in order to create a unique atmosphere. which has to be sophisticated and easy going at the same time.

魅惑

The quality of this space is expressed by every detail, and also by the use of different materials.

质感的表现，在于每个细节之处，也在于不同材料的运用上。

在此项目设计之初，设计师需要为整个设计的风格定义。最终，设计师发现“经典现代风”是对这个空间的最好诠释。Drop酒吧所在地是上海闹市区为数不多的大厦集中区，也是为数不多的拥有原创天花、壁面设计的空间。它原本就拥有它自己的特色和灵魂。

根据OOBIQ建筑设计团队给它的定义——“经典现代风”，就是要在室内达成一种融合经典材质、传统造型和材料于一体的设计，这样才能符合像上海一样的国际大都市的身份。整个俱乐部都采用同样的材质装点：几何形状的摩洛哥式瓷砖成为大理石地板的珍贵装饰品。这种暖色调装饰物成为Drop酒吧的另一元素——不同的文化与时空联系在一起，就能够创造一种独特的氛围，一种既复杂又简单的氛围。

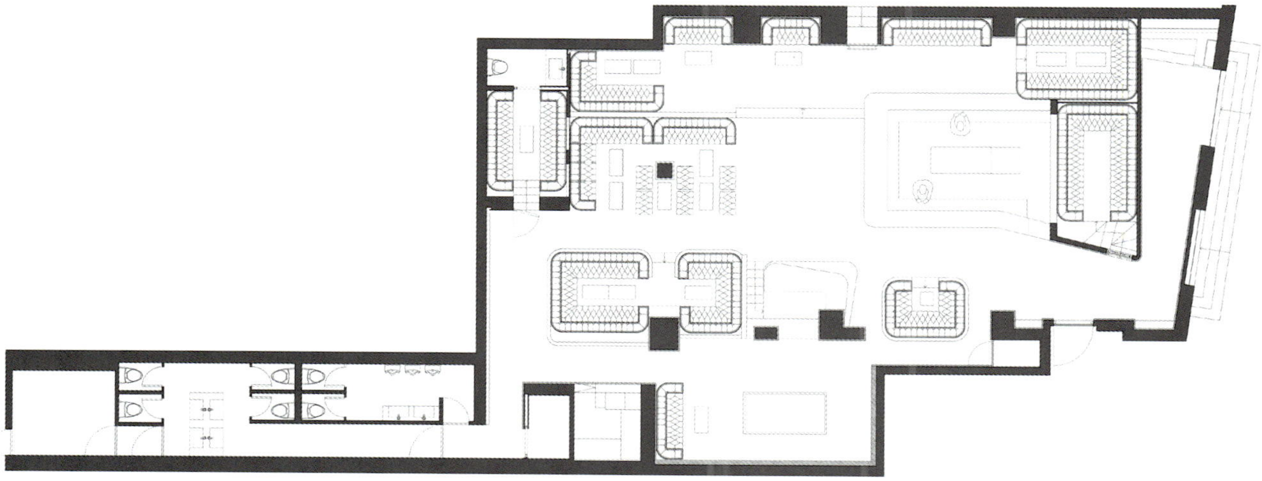

公元一号酒吧

The No.1 Bar

设计师：曾建龙
设计公司：NOTA香港格瑞龙国际设计有限公司
项目地点：中国温州
项目面积：600平方米
主要材料：大理石、不锈钢、玻璃镜子等

The No.1 Bar is a very special entertainment space of bar. Firstly, its geopolitical location is of special ground surface properties, which resides at a cave in a hot city park-like plaza, and in peace time its outside is a plaza available for entertainment of common people. It was very inspiring when the designer received this project, because it was very funny, then the positioning of the design is required to be shining and of strong fashionable sense, and at the same time, the nature of space is displayed by contrast and theatricality in the design.

魅惑

Colorful elements, abundant formations and simple design ways highlight the unique flavor of the space.

多姿多彩的元素，丰富的造型，化繁为简的手法，突显出空间的独特品位。

Pioneer

公元一号是一个比较特别的酒吧娱乐空间，首先它的地理位置具有特殊的地标性质，它处在一个比较热闹的公园广场的山洞里，外面是供老百姓进行娱乐活动的广场。当接到这个项目的时候设计师是很有激情的，因为觉得好玩，所以一开始在设计定位上就要求比较炫，时尚感强，同时设计以对比性和戏剧性的方式去呈现空间本质的特性。

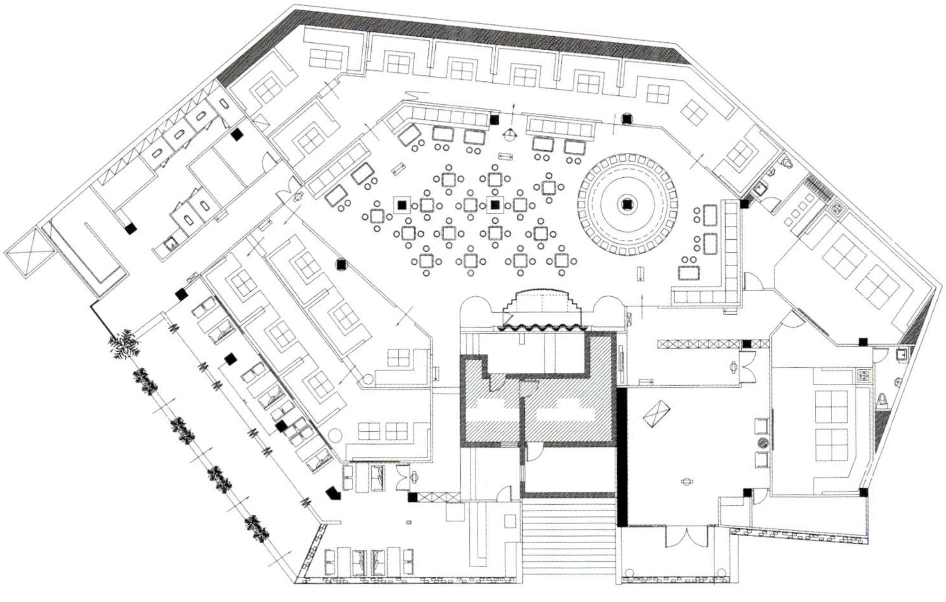

苏州皇家TT酒吧

Suzhou Royal TT Bar

设 计 师：孙亮
设计公司：成都优佰文化传播有限公司
项目地点：中国苏州
项目面积：600 平方米
主要材料：实木、茶镜、教堂玻璃、铁艺等
项目摄影：香港ATA影业有限公司

The bar-popular wooden structure is adopted into this case, which is well combined with Suzhou City. Wooden structural material can be the great integration of the city's cultural characteristics and the space. The use of the lighting becomes particularly important when the main material reflects both the original color and texture and the funny atmosphere. In order to be controlled on-site at any time, the lights of this case are not functional light source, but a group of controlled lights. In terms of the color, the bar is mainly dressed with red, yellow and other warm colors,dotted with blue and white, creating a warm and comfortable atmosphere. Key considerations were given to the use of the material from the point of the material itself instead of deliberately changing its texture. The video wall also becomes an important tool to adjust the color and atmosphere, increasing the dynamic and making up for the lack of color according to the change of the lighting.

魅惑

The attractive things are not only about the appearance, but also about the flavor of the city itself.

魅惑你的不仅仅是其华丽的外表，还有这座城市本身所具有的浓烈的韵味。

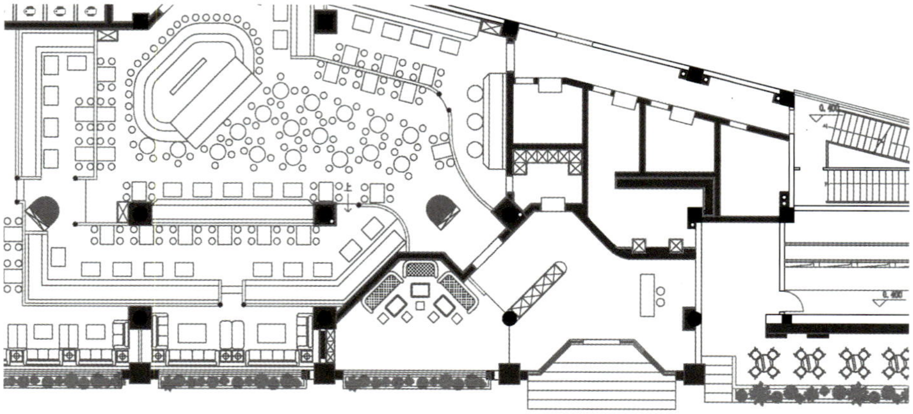

本案采用目前酒吧比较盛行的木制结构，这首先与苏州这座城市很好地结合起来了，木质结构的材料能够很好地将这座城市的文化特质融入空间之中。主要材料要体现比较原始的色彩和质感，同时又要能体现出娱乐的氛围，所以灯光的运用就显得尤为重要了。酒吧的光线要暧昧，明暗对比相对较大，所以本案中所有的灯光都不是泛光源，且全部由调光台控制，以便于现场随时调控。色彩上，主要以红、黄等暖色为主，局部点缀蓝色、白色，以营造温暖舒适的氛围。材料的运用上，重点考虑材料本身的特点，不刻意改变其质感。视频墙也是调节色彩和氛围的重要元素，根据不同时段，随着灯光的变化而变化，不仅可以增加动感，还可以弥补色彩的缺失。

新加坡克拉克码头诊所酒吧

The Clinic at Clarke Quay Singapore

设计公司：Concrete Architectural Associates
项目地点：新加坡
项目面积：1200 平方米

Clinic is the medicine, which brings you to a new dimension of partying. The floor plan is built up out of different pills (medicines), each with its own form, ambiance and experience or side-effect. Clinic consists roughly of two parts: The club is located downstairs, with different pills (spaces) named after medicines. The Vip-lounges and restaurant are upstairs, again with a variety of pills but named after states of mind. As you enter the emergency admission room, a round, clean, white aspirin pill with a red cross on the floor will come to you. The break-line of the pill provides indirect light from the ceiling.

魅惑

The space has a kind of magic power, where every detail could bring people into another world.

空间有种撩人心弦的魔力，每个细节都可以将顾客带入一个新的境界。

EXIT

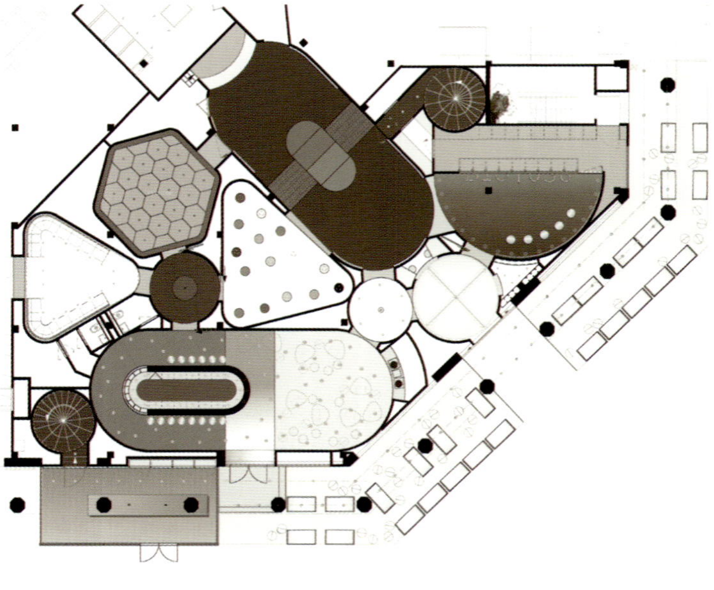

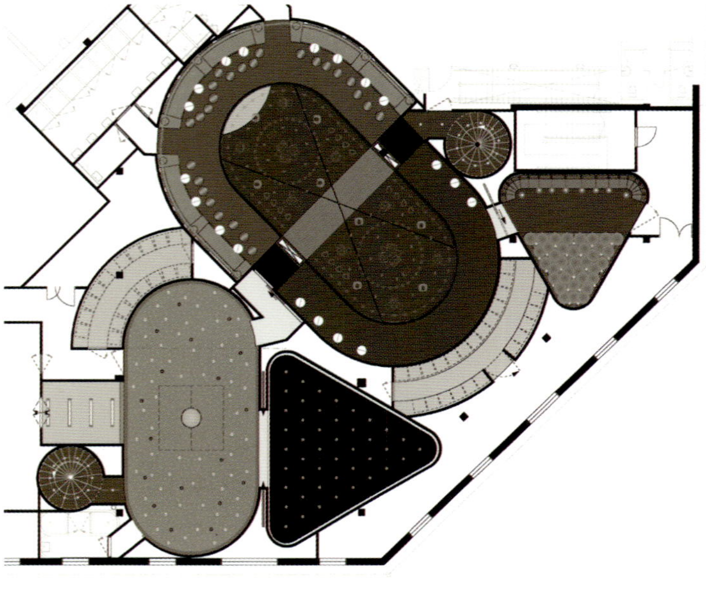

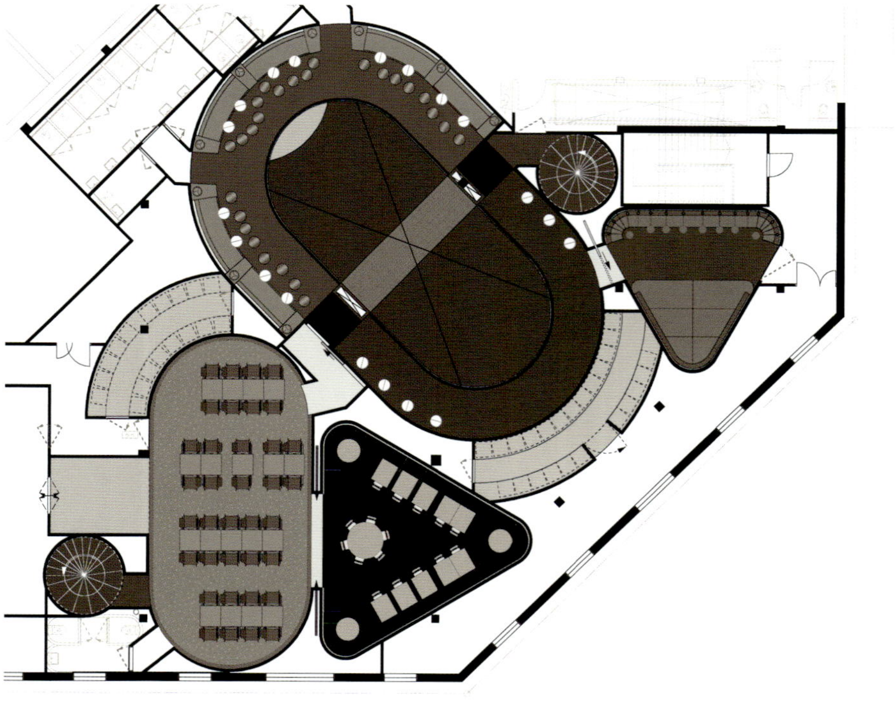

诊所酒吧形同药物，把人们带到派对新境界。平面布局由不同药丸形状的包房组成，各有各的形态、氛围体验或副作用。诊所酒吧主要包括两部分：俱乐部在楼下，这里的不同空间均以药物命名；VIP休闲吧和餐厅在楼上，房间又是各式各样的药丸胶囊形状，但以不同的精神状态命名。进入急诊住院室，映入眼帘的是一个圆形、洁净、白色、形同阿斯匹林药片的房间，地板上有一个大红“十”字，而天花板上的十字镂空为空间提供了间接照明。

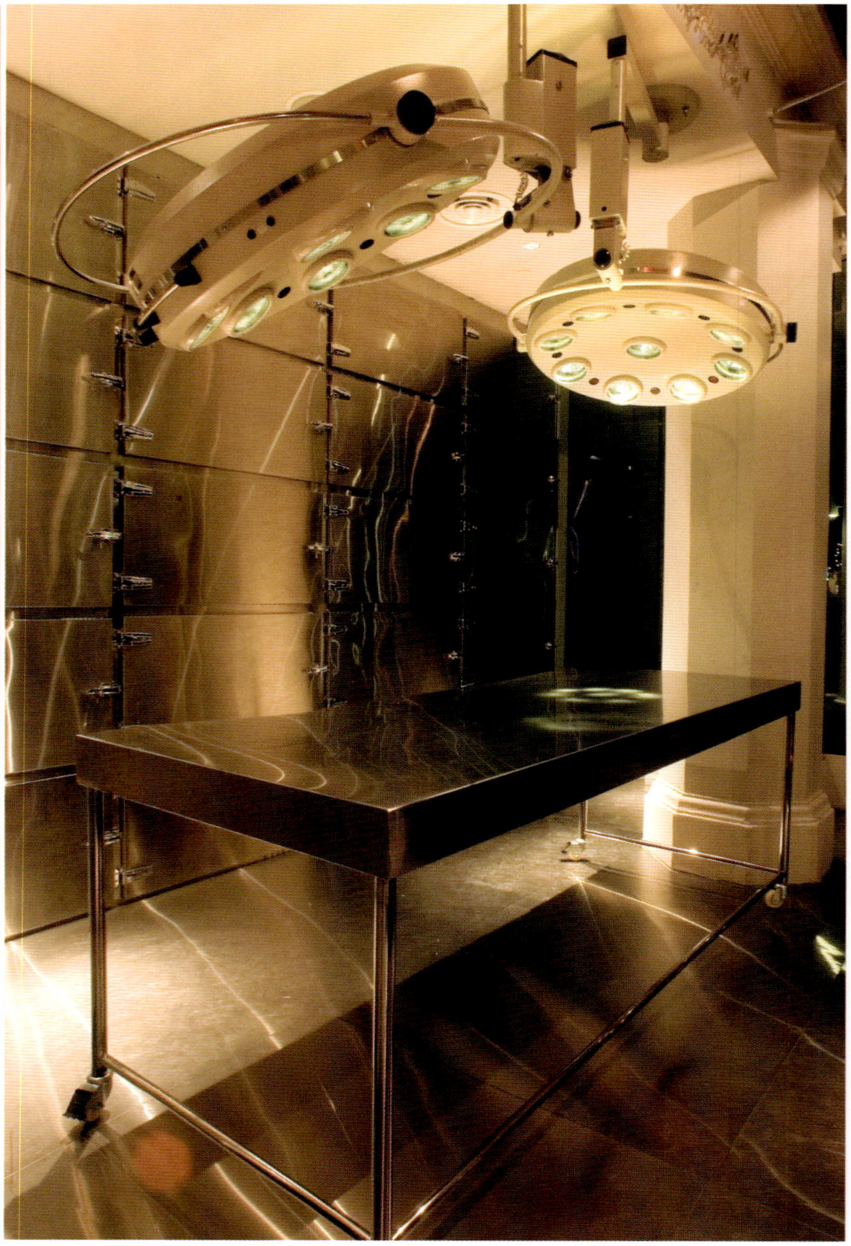

Au
Gold

Press Club都市红酒吧

Press Club Urban Wine Venue

设计公司: BCV 建筑事务所
项目面积: 762平方米

The charge for this innovative wine collective/tasting lounge was to create a sophisticated urban identity while simultaneously realizing a modern expression of the spirit of California wine country. Within a clean-lined, minimalist aesthetic, the architects have captured the casual sophistication of the wine country and created a space that is both timely and timeless. Taking cues from Napa's unique blend of industry and natural beauty, the restrained design celebrates the juxtaposition of the industrial against the organic as a foundation for the projects expression.

A sense of place has been created within an otherwise challenging urban location. The majority of the project's 8,200 sf is housed within the subterranean cellar. The cellar program includes individually-maintained tasting bars for 8 regional world-class wineries, public and semi-private lounge spaces, a central wine and food bar, a private event room and support spaces.

A bold monolithic concrete stair dramatically links the upper level arrival and wine shop to the extensive wine-centric area below, flanked by a 17-foot-tall LED back-illuminated bottle wall. This signature feature wall transcends the divide between entry level and cellar and punctuates the project with its striking color patterning.

魅惑

Based on elegant and free style, the designer wants to create a fashionable yet classical space.

以酒吧随意、优雅的特质为表现重点，构建一个既时尚又经典的空间。

PRESS CLUB
EXIT

这个新颖的藏酒、品酒吧旨在打造一种高级的都市身份，同时以现代表情来呈现加利福尼亚酒庄的精神。

在线条干净、简约的美感里，建筑师捕捉到了酒庄随意又优雅的特质，构建了一个既时尚又经典的空间。Napa酒乡融合了工业和自然的独特美感，给了建筑师启发和灵感，而内敛的设计正是向这种工业化跟有机生态的相辅相成致敬，两种并列存在的元素成为项目表达的基石。

该设计在一个相当具有挑战性的都市位置创造了场所感。762平方米的项目主体位于一间地下酒窖。酒窖包括供8家世界级的区域性酿酒商、单独维护的品酒吧，公共和半私隐性质的大堂吧空间，一个中央酒水、食物吧，一个私人活动室和配套空间。

气势恢弘的单块巨石混凝土阶梯连接了上层入口、红酒店铺和下层宽广的酒类集中区，阶梯侧墙是一面17英尺（约5米）高的LED背光酒瓶墙。这个标志性的特色墙超越了作为入口层跟酒窖之间的隔墙的意义，用其醒目的色彩派绘画作品为该项目画上圆满的句号。

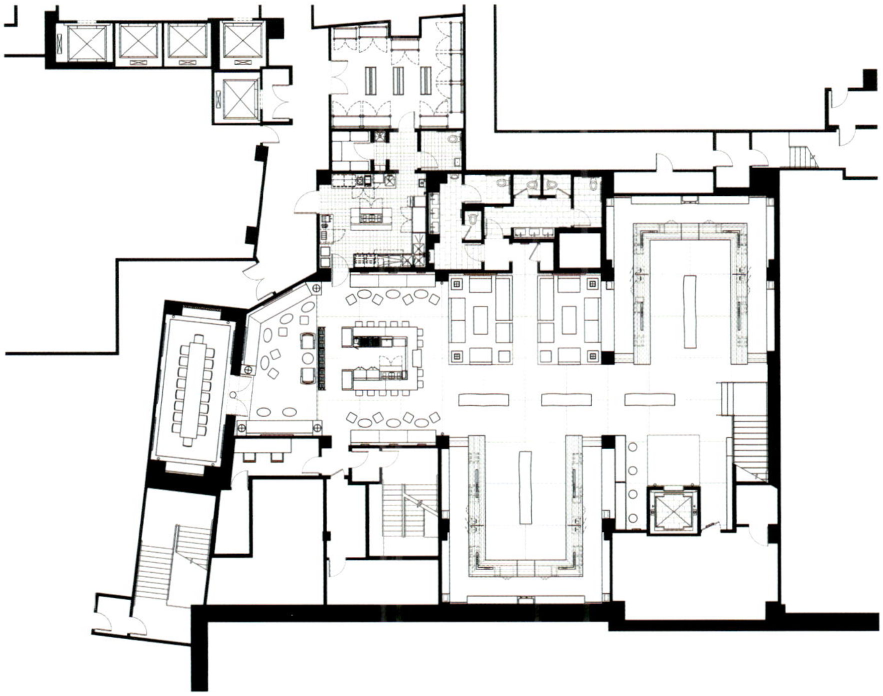

G-ZERO俱乐部

G-ZERO

设 计 师：Duilio Damilano
项目地点：意大利皮埃蒙特
项目面积：175平方米
项目摄影：Andrea Martiradonna

The restaurant, located in a famous ski location in Piedmont, has many constraints in form and structure.

In particular, the network of pillars was masked by the realization of "mushroom" shaped wood and enamel that have become the distinguishing feature of the room.An eye behind the counter brings light to the rear kitchen and masks other two pillars. The black ceiling tends to zero by surfacing only sources of light, sort of flying saucer floating above their heads. A wall of natural stone stretches up into the bathroom, in contrast with the angle of crystal above the entrance to washbasins. Serigraphs which include the conical shapes of the "mushrooms" filter the view to the bathroom. The wooden floor warms the entrance area, and turns to stone, to draw mountain atmosphere.

魅惑

Located in a famous ski location in Piedmont, the imaginative design is favored by lots of people, and the use of materials expresses its strong natural feeling.

该俱乐部在著名的皮埃蒙特滑雪胜地，其富于想象力的造型深得人心。材质的运用表现出了强烈的自然感。

zero

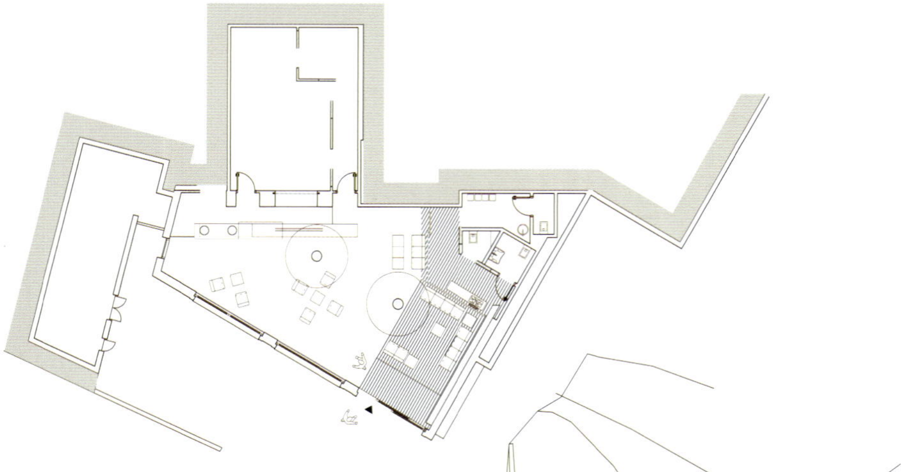

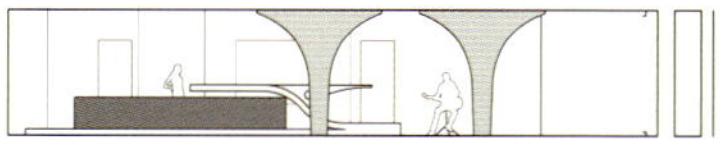

这个餐厅俱乐部坐落在著名的皮埃蒙特滑雪胜地，其设计在形式和造型上有许多限制。散布的支柱以蘑菇的造型出现，变成了房间里面颇具特色的设计。黑色天花板只露出了几缕光，灯光下是茶托似的造型设计。用天然石头打造的墙面一直延伸到洗手间，与洗手池入口处的水晶斜角形成鲜明的对比。洗手间也运用了绢网印花似的圆锥体造型，用“蘑菇屋”造型把视野过渡到洗手间里面。木质地板增强了入口的温馨感，然后转变为天然石材地板，并引入了自然的山间氛围。

喜力酒廊

The Heineken Lounge

设计公司：UXUS Design
项目地点：美国新泽西
项目面积：156平方米
项目摄影：Dim Balsem

A bold new bar concept designed for International Airports brings the Heineken brand to life, for travelers the world over. It is an inclusive experience, giving travelers a "third place" to escape the chaos and stress of international airports and offering the benefit of Heineken's exclusive Extra Cold chilled beer system, a fully stocked cocktail bar, within the relaxing environment you would expect from a premium airline club. The environment is cosmopolitan, designed with international travelers in mind. It incorporates Dutch design icons with custom designed seating areas created to help relax and refresh travelers in their transit. Special Cabana seating offers a sense of privacy in a public space with the comfort of a luxurious leather sofa. Heineken green anodized aluminum tables, Wengé wood room dividers, and atmospheric lighting, all contribute to the sophisticated Heineken Lounge experience.

魅惑

The originality is the feature of bar design. There is no noise, no pressure but only quiet atmosphere inside the bar which give the absolutely relaxing air to each person.

富有创意是酒吧的特色所在，这里没有喧嚣，也没有压力，沉静的空间之中，给予每个人 的是一份绝对的休闲气息。

Heineken
LOUNGE
Heineken
LOUNGE

Heineken
LOUNGE
Heineken
LOUNGE

Heineken
LOUNGE

Heineken
LOUNGE
Heineken
LOUNGE

Heineken

Heineken
LOUNGE
Heineken
LOUNGE
HEINEKEN PREMIUM LIGHT

HEINEKEN LAGER BEER
Heineken
BREWED IN HOLLAND

为国际机场设计的喜力酒廊采用了大胆创新的酒吧概念，让世界各地的旅客体验到喜力啤酒品牌的活力。丰富、兼容的体验给了旅客一个“第三地”以逃避国际机场的喧嚣、混乱、压力，这里有喜力独一无二的特冻冰啤酒系统、一个存酒齐全的鸡尾酒酒吧以及高级航空俱乐部应有的轻松环境。

极富大都会特色的环境在设计时考虑了国际旅客的需求。荷兰设计符号跟定制座椅区相结合，有助于让转机的旅客放松心情，恢复精力。特别的开放式小屋座椅给公共空间带来一种私密感，同时豪华皮革沙发让人倍感舒适。喜力绿色电镀铝桌、铁刀木隔断和营造气氛的灯光照明全都致力于呈现高雅的喜力酒廊。

Spirito俱乐部

Spirito

设计师: Puresang
项目地点：比利时布鲁塞尔
项目摄影：Mark Groeneveld

When Francesco Ravo had the chance to buy this beautiful Anglican church in the Rue de Stassart, he didn't hesitate for a second to do something with it. How many times in a lifetime do you get the chance to give a beautiful, historical church a total makeover? Francesco asked Puresang to help him. Together they made a very big and classy hospitality-project. Thanks to the many years of experience in those kinds of projects, Puresang could make an absolutely superb project of it. The concept is a mix of an event location, a lounge bar , and a cosy club at the weekends. The intention was to create a very high level environment for international public. The interior is a mix where "modern glamour meets gothic extravagance". The project contains 3 bars with 5 different lounges and a private room. People should be able to enjoy the luxurious top class champagnes, divine liqueurs or a delicious Martini on the heavenly beats of a liveband or DJ.

魅惑

The interior is a mix where "modern glamour meets gothic extravagance", and in order to create mutil-function space, the visual sense and materials are perfectly presented to us.

室内设计集摩登魅力与哥特式奢华于一体，打造多功能空间，在视觉以及材料表现上显得游刃有余。

MARTINI
BACARDI
RESERVA
WILLIAM LAWSON'S
JACK DANIEL'S
Tennessee
WHISKEY

当Francesco Ravo有机会购买这个坐落在Rue de Stassart的美丽教堂Anglican church时，他没有丝毫犹豫。在人的一生中能有多少机会把一个美丽的、历史悠久的教堂重新翻新？Francesco找到了Puresang帮助他完成整个设计工作。他们一起完成了一个艰巨且宏大的项目。得益于这些年来项目设计的经验，Puresang已经能够设计出极好的项目品质。本项目的设计理念是设计一个集会议室、高级酒吧、舒适会所于一体的空间，主要是为国际性的聚会提供一个高级的集会场所。室内设计集摩登魅力与哥特式奢华于一体。项目包含3个酒吧，有5个不同的休息室和一个贵宾间。人们能够在DJ或者现场乐队的节奏中，品尝到最奢华的香槟、顶级的利口酒，或者美味的马提尼。

圣地亚哥酒吧

Bar of Santiago

设计师：辛军
设计公司：深圳市辛视设计有限公司
项目地点：中国山西
项目面积： 500平方米

Bar of Santiago is with some European neoclassical flavor, which also has some fashionable luxury flavor. The space has many space dimensions and it is also divided into five individual lounge areas - the beautiful and relaxing bed lounge, the luxurious castle lounge and soft party lounge, etc. The feeling is different when people are on the different areas.

The colors of the bar and different elements create the European neoclassical flavor which makes people feel fresh, and the strong relaxing flavor of the bar is the divine place of people's soul. The designer adds some fashionable elements, so the bar appears more international and diversified.

魅惑

The design merges European classic and modern fashion which has more free space for expression, and the space becomes more cozy.

集中了欧式古典与现代时尚，表达上更加自由，空间上更加舒适。

圣地亚哥酒吧带有一点欧式新古典韵味，又夹着一些时尚的奢华。有原地平、错层、夹层多个空间层次。分为5个独立卡座区：香艳休闲的床式卡座区、豪华的古堡式卡座区、柔美的派对卡座区等，坐在不同区域的感觉是不一样的。

由酒吧色调和不同的元素打造的欧式新古典韵味让人眼前一亮，浓郁的酒吧氛围让酒吧本身成为心灵的圣地。另外利用一些时尚元素，使得酒吧的多元化特点更加明显。

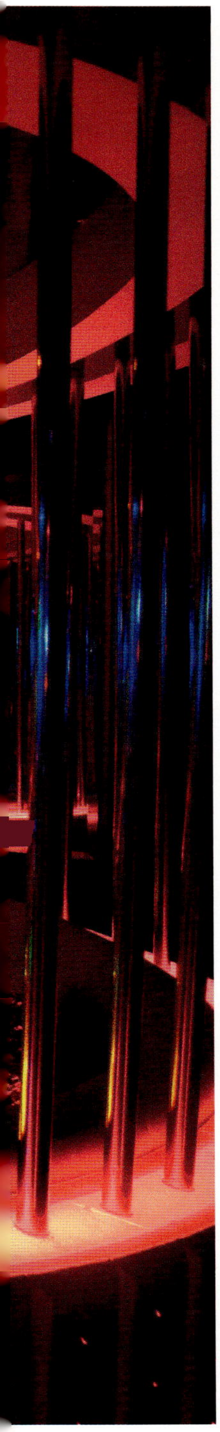

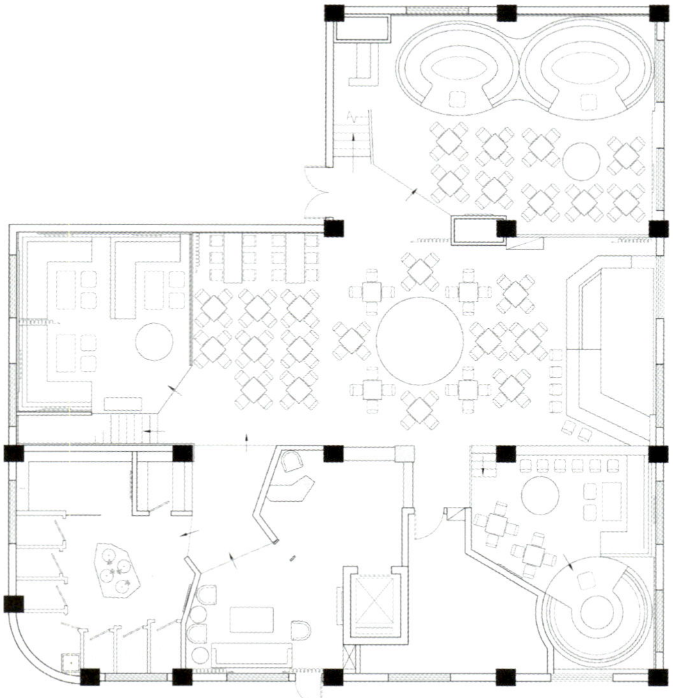

Show吧

Show Bar

设计师：吴伟宏
设计公司：厦门宏盟东方室内设计机构
项目地点：中国厦门
项目面积：450平方米
主要材料：皮革硬包、不锈钢、光纤等

The design focuses on Black & White, which shows a clean and sharp environment. The designer makes it fashionable by using modern furniture. The whole space is of geometric plane and lines. The designer connects the whole space by network, and integrates the space together. With the projection and lighting, it creates a vivid, warm atmosphere. The baking finish iron plate floor is dark gray, showing the designer's pursuit of space retroaction.

魅惑

The designer uses "black and white" to create the space, however, there are other colors inside the space. The most important is that the designer wants to find a good way to express and highlight the black and white theme.

设计师以“黑白”为主题打造空间，但是实际上，这里并非只有黑与白。重要的是，设计师找到了一种很好地表现和突出黑白主题的方式。

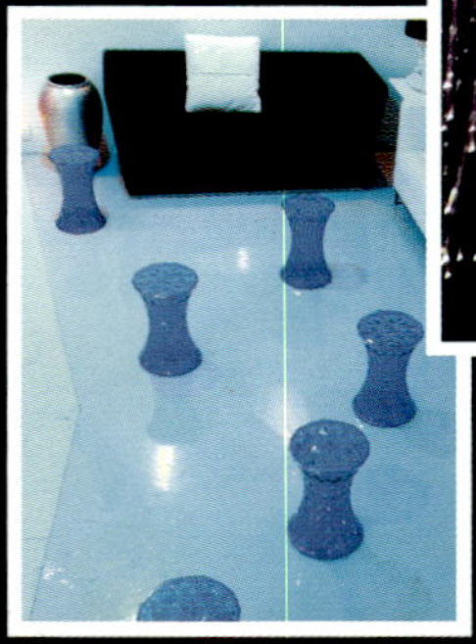

本案以“黑白”为主题，整个室内空间强调整体的纯净和利落感。运用现代家具元素，表达了设计者要求现代、时尚的意图，整个空间以几何面和线来组合，在墙面添加了多媒体设备。整个空间以网状的形式连接起来，形成一个整体。投影与灯光营造出了一种活泼热情的气氛。烤漆钢地板是暗灰色的，表达设计者追求空间逆动的意境。

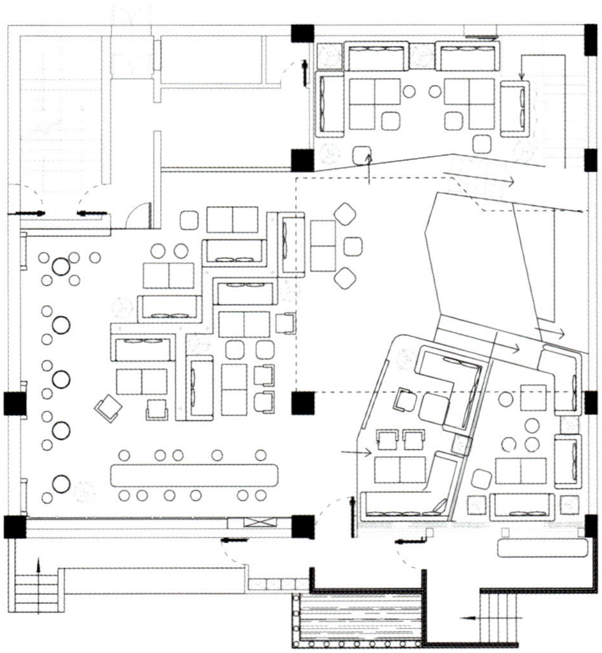

一层

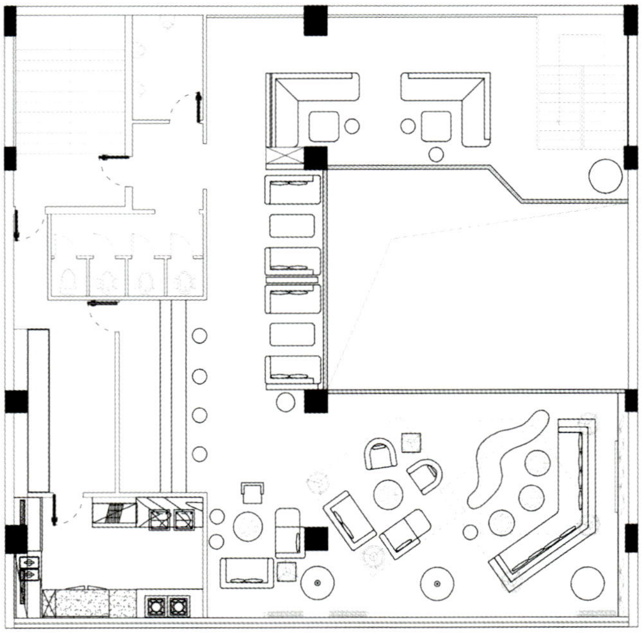

二层

上海拉雷红酒吧

La Lé Red Wine Bar in Shanghai

设计师：林琮然
设计公司：CROX阔合国际有限公司
项目地点：中国上海

This project is located at Yandang Road of Shanghai. The designer uses his fashionable design language to express his design and to create a space where even the city of Shanghai could indulge and release itself. Next to the Fuxing Park, there are lots of cirrus. Through firmiana formation, there is a forest which creates the bar area, drinking area, wine cellar and private room. The different spaces and gestures magically change the people's mood and harmoniously combine the La Lé bar and people's fun time. The modern yet futuristic space adopts marble and glass to create high quality feeling, the design of the ceiling is full of fluent line feeling, and the wooden material of the wall looks like the musical note. The curveed mirror surface passageway subtly expresses its soft flavor, and it looks like the blurry scenery which you can't see clearly.

魅惑

The fashionable sign and natural design language highlight the space connotation and quality – cozy and relaxing.

时尚的符号以及自然、前卫的设计语言，突显出了空间的内涵与品质：舒适，尽情释放天性。

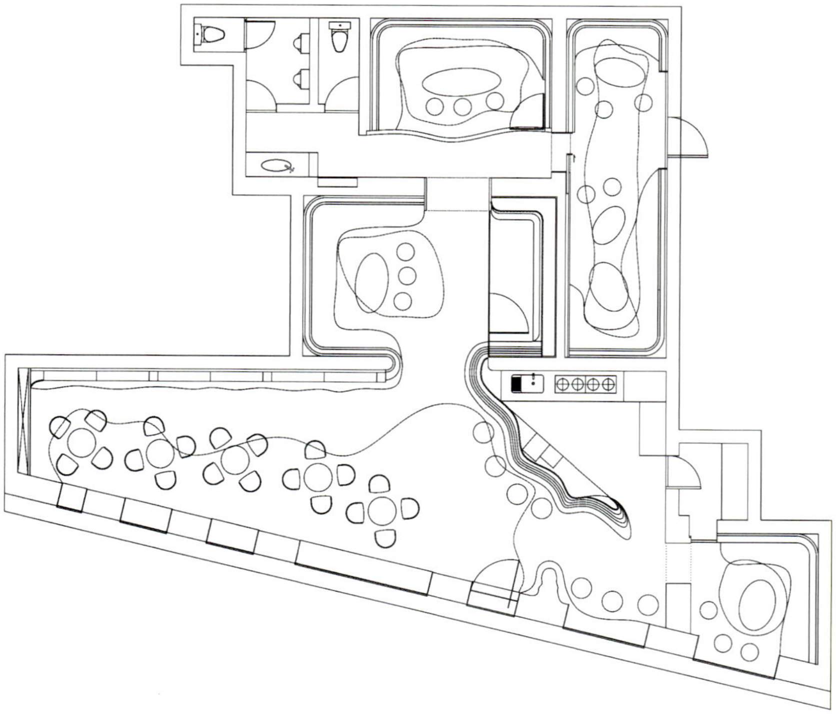

该项目位于上海最早的休闲步行街雁荡路上。设计师用他前卫自然的设计语言去演绎、打造一个让挑剔的上海人也愿意沉溺和释放的人性空间。邻近遍布梧桐树荫的复兴公园是上海记忆的藤蔓，La Lé 空间透过如梧桐树干的结构柱逐渐展开，形成一整片树林，创造出吧台区、试饮区、酒窖及独立包厢。不同的空间与坐姿幻化了人们的情绪，而La Lé 红酒吧与人们交融出饮酒空间那自然且美好的时光。既现代又未来的空间采用大理石及玻璃打造质感， 而天花空间充满流线、动感，墙面木材如音符般排列出美妙的韵律，弧形镜面不锈钢板通道巧妙诠释微醺情绪，增添柔和韵味，如看不透的景色，展现柳暗花明又一村的感动。

碳吧

The Carbon Bar

设计公司：B3 Designers
项目地点：英国伦敦
项目摄影：Sean Pines

The Carbon Bar was the new addition to the Cumberland Hotel, which also houses Gary Rhodes' W1 restaurant, designed by Kelly Hoppen. The bar is based along Old Quebec street, just off Oxford Street. The interiors inspiration comes from the Shoreditch style industrial architecture. B3 Designers was approached by the Guoman Group to design a new bar.

The general materials used to create this industrial interior were a combination of concrete, brick, steel, mesh and leather. This interior was designed to increase the amount of space available for the clients has private areas but at the same time has everything visible.

The bar has been designed so that it has descriptive words that would be found in industrial sites. The bar itself is 14 metres long, providing a vast amount of space for people to be able to get drinks from, along with a mezzanine floor, suspended above the lower floor which has a champagne bar. When going upstairs, there is a featured wall of champagne, which is two stories high, which is filled with some of Taittinger's most expensive and rare bottles. Another feature wall has barrels along it adding to the industrial look.

There is a chain room that is on a raised floor, surrounded by huge steel chains hanging from the ceiling to the floor which hosts the VIP area, within which is a bar so that the VIP's can stay within that area.

魅惑

The building affords the inspiration to the designer. The elegant arrangement and good visual effect show the humanistic service of the bar.

工业建筑给设计师带来了一些灵感，优雅的格局和良好的视觉效果表现出了酒吧的人文关怀。

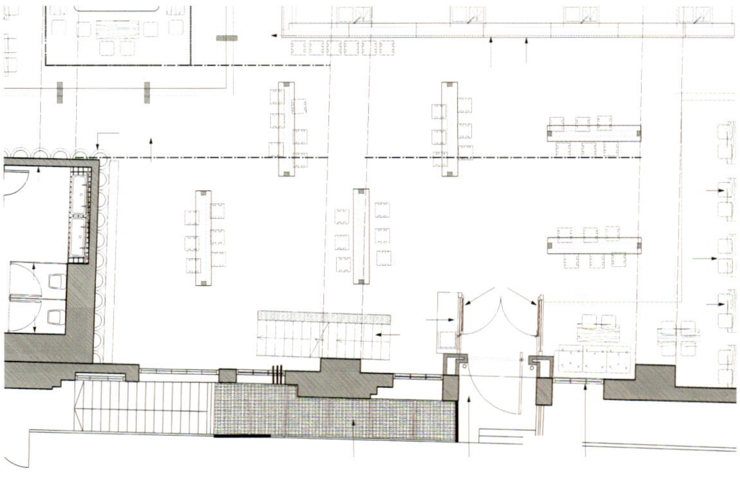

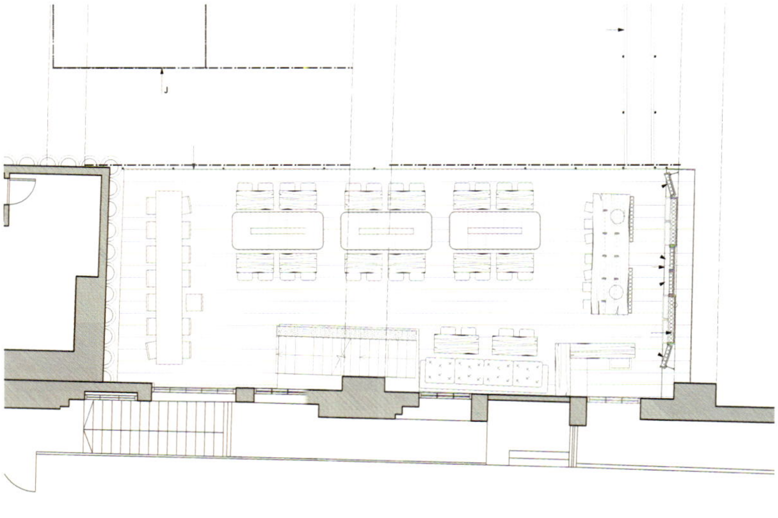

HAZARD
SOUND
EXIT BEWARE: HEAVY MACHINERY IN
PROTECTIVE GEAR EXIT BEWARE: HEAVY
VIBES WEAR PROTECTIVE GEAR EXIT

碳吧是Cumberland酒店的新增项目，加里·罗德大厨的W1餐馆（由Kelly Hoppen设计）也在该酒店内。酒吧位于老魁北克街，就在牛津街旁边。内饰灵感来自东伦敦的新兴文艺区风格的工业建筑。Guoman集团委托B3 Designers设计公司为其旗下酒店设计一个新酒吧。

营造工业化内饰效果的材料是混凝土、砖、钢、铁网和皮革。碳吧的设计使其空间最大化，拥有私密空间，同时又能让人一眼尽览其全貌。

碳吧具有工业基地的建筑特色。这里有一个14米长的大酒吧，为人们提供饮酒的巨大空间，还有一个悬在低层地板上方的夹层香槟酒吧。客人可经楼梯上到这个夹层空间。楼梯毗邻一个二层高的香槟酒墙，里面摆满了泰亭哲酒庄一些最昂贵、稀有的香槟酒瓶。另外一个特色墙堆满了酿酒桶，增添了工业化的感觉。

抬高的地面上有一个链条屋，从天花板悬垂至地面的巨型、固定式钢链把这里围合成贵宾区，这些链条在主吧区内又营造了一个独立屋的效果。

+x酒吧

+X Bar

设计师：Dimitris Naoumis
项目地点：希腊纳基

His creative ideas and inspiration gave a unique character to '+x', considering that each one of them has already important creative routes in the contemporary art and design life of Greece. "+x" is based on elegant underground design forms, and this uncommon combination gives the place its own special aesthetic character.

Some of the main design elements are the combination of warm and cold colors, and the mixture of the street art references and the inspiration from Avant Garde art movements. These elements are mixed together harmonically. The creators gave special attention to every detail and most of the furniture and the lighting installations are designed especially for the place.

魅惑

The classic elements adds the artistic feeling of the space which creates the unique visual aesthetic sense.

古典元素的运用增强了空间的艺术气息，从而营造出了独一无二的视觉美感。

GREATEST
MEDITATION
IS
A
LETS
GO

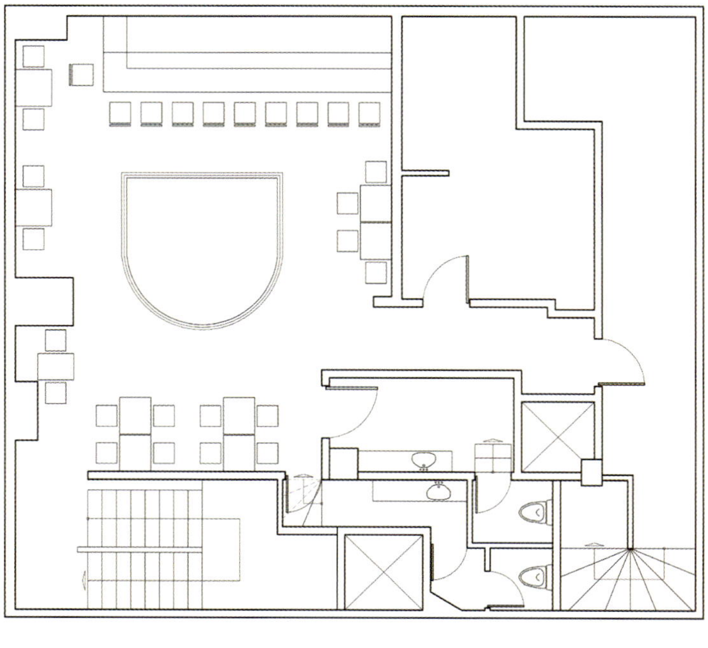

设计师充满创造力的构思以及灵感为“+X”提供了独特的品质。“+X”的设计是根据典雅的地下室设计形式设计而成的，这个不一般的结合让这个地方有其特殊的美感。

一些主要的设计都是冷色调与暖色调相结合而产生的，是街头艺术与先锋派艺术运动的灵感相融合的产物。这些元素都很好地结合在一起，并且设计者还特别注重空间的每一个细节和绝大部分的陈设，还包括为此空间特殊定制的灯光装置。

梧桐红酒坊

Wutonghong Venue

设计师：李景光
设计公司：上海景光室内空间设计创意坊
项目地点：中国上海
建筑面积：890平方米
主要材料：三色仿古砖、黄洞石、白洞石、黑镜、壁纸、黑檀、银箔漆、雅士白大理石、镜面马赛克等

In Jingan district of Shanghai where row upon row of bars and cafes exist, it might not attract passers–by's much attention if another bar appears in such a prosperous, luxurious environment with the scene of debauchery, however, Phoenix Tree Wine Venue presents itself in public sharply, whose special air that is simple and unsophisticated yet modern and cozy indeed appeals to a large number of people favoring red wines.

Dark materials are used remarkably in the project, in which the designer planted mirrors and shiny metal colors. Combination and match of various colors and lights, plus personalized artistic treatment, produce a more layering visual sense, deeply affecting people and making them fond of the environment immediately. Rosewood in coffee or black dominating the space completely creates a classical, heavy feel, but the resulted deep spatial environment is cleverly enlivened by bright colors of metal, mirrors and mosaic embedded in the space, thus, the special, flowing effect which is classical yet rich in fresh, stylish elements is achieved! Every piece of carefully chosen sofa, chair, bar stool and gorgeous colors infuse an art atmosphere into the interior space, which, under the influence of lights, looks more magnificent, distinguishing the bar from all others.

魅惑

The classic feeling makes people would like to stay at here. The coffee color and red sandal wood highlight the classic and steady quality of the space.

古典与厚重感让人们更加乐于在此驻足，咖啡色以及紫檀木等大大提升了空间的古典、厚重气质。

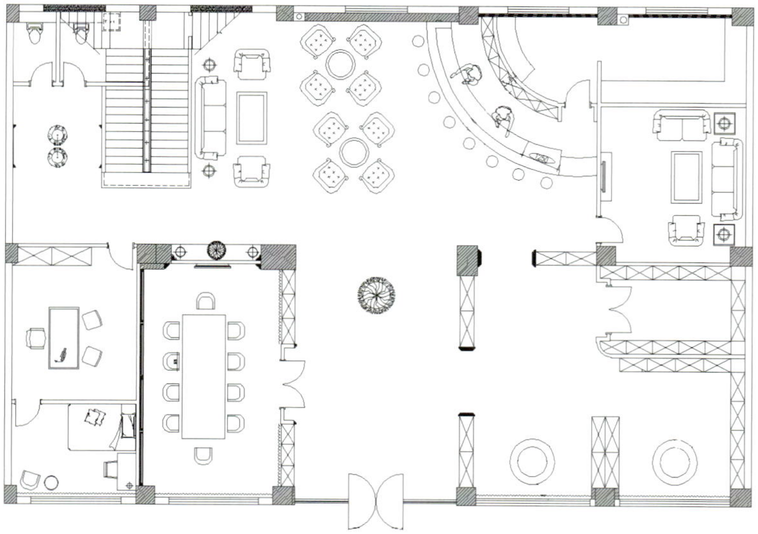

一层

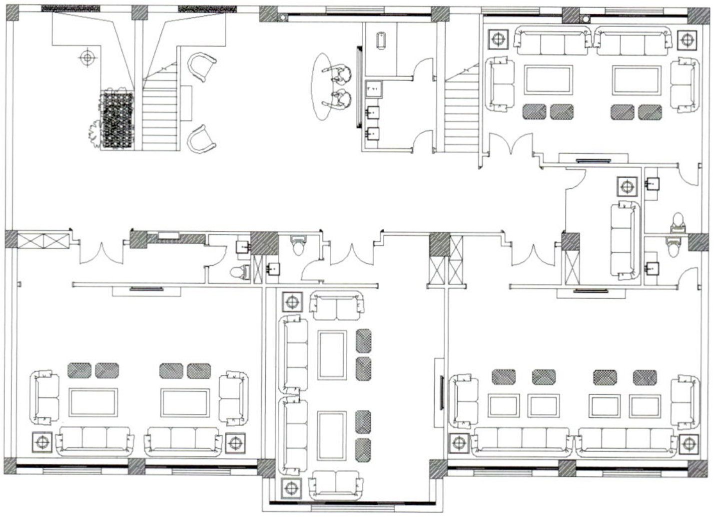

二层

在上海静安区，酒吧、咖啡屋可谓鳞次栉比，在这样一个灯红酒绿、繁华豪奢的环境里多一个酒吧店面，应该是不会引起路人很多注意的，但梧桐红酒坊的出现却让人眼前一亮，其既古朴优雅又现代温馨的特有风情着实吸引了一大批红酒客。

本案设计师多采用深色材料，并穿插镜面及亮金属色于其中，通过不同色彩及光线的组合搭配，再加以个性化的艺术处理，令视觉效果更有层次感，让来到这里的人

深受感染，并很快喜欢上这里的环境。本案以精选深咖啡色或黑色檀木为主，充分营造出了空间氛围的古典、厚重感，但由此造成的深沉的空间环境又被嵌入其中的金属、镜面、马赛克的鲜明色彩巧妙地激活了，从而收到特有的既古典又富有新鲜时尚元素流动的效果。各种沙发、椅子、吧凳都经过了精心的挑选，加上斑斓的色彩，令空间的内部环境更加充满艺术气息，在灯光的作用下更加美轮美奂，让整个酒坊独树一帜。

图书在版编目（CIP）数据

魅惑夜生活 / 徐宾宾 编. —南京：江苏人民出版社，2011.5
ISBN 978-7-214-06988-7
Ⅰ. ①魅… Ⅱ. ①徐… Ⅲ. ①文娱活动－文化建筑－建筑设计－世界 Ⅳ. ①TU242.4

中国版本图书馆CIP数据核字(2011)第058378号

魅惑夜生活

徐宾宾 编

责任编辑 刘 焱 赵 萌
版式设计 金陈晨
文字编辑 余 京 张诗静
责任监印 马 琳
出　　版 江苏人民出版社
发　　行 天津凤凰空间文化传媒有限公司
销售电话 022-87893668
网　　址 http://www.ifengspace.cn
集团地址 凤凰出版传媒集团（南京湖南路1号A楼 邮编：210009）
经　　销 全国新华书店
印　　刷 当纳利印刷（深圳）有限公司
开　　本 1000mm×1270mm 1/16
印　　张 19
字　　数 152千字
版　　次 2011年5月第1版
印　　次 2011年5月第1次印刷
书　　号 ISBN 978-7-214-06988-7
定　　价 288.00元(USD 58.00)

（本书若有印装质量问题，请向发行公司调换）